W9-CDL-655

Popular Complete Smart Series

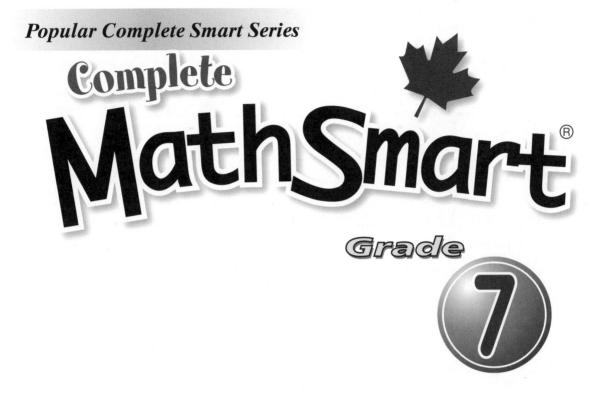

Complete
MathSmart

Grade
7

Proud Sponsor of the Math Team of Canada 2017

ISBN: 978-1-897164-21-1

Contents

ISBN: 978-1-897164-21-1

* The Canadian penny is no longer in circulation. It is used in the units to show money amounts to the cent.

ISBN: 978-1-897164-21-1

ISBN: 978-1-897164-21-1

Overview

In this section, Grade 7 students expand on previously developed skills with fractions, decimals, and percents learned from Grade 6. They will also advance into higher mathematical concepts such as exponents, integers, and algebra.

Students will learn to relate fractions to decimals, percents, ratios, and rates. There is a heavier emphasis on applying the correct order of operations in evaluating various mathematical expressions.

New skills include evaluating exponents, squares, and square roots. Students will work with integers and learn algebraic concepts such as collecting like terms, writing, and solving equations.

Number Theory

1. $3 \times 3 \times 3 \times 3 = 3^4$ ←— 3^4 is a power; 3 is a base; 4 is an exponent

    $= 81$ ←— 81 is the value

2. Comparing powers:

    a. $4^7, 4^5$ ←— since the bases are the same, compare their exponents; the greater the exponent, the greater the value

    $4^7 > 4^5$

    b. $3^9, 6^9$ ←— since the exponents are the same, compare their bases; the greater the base, the greater the value

    $3^9 < 6^9$

    c. $3^4, 4^3$

    $3^4 = 3 \times 3 \times 3 \times 3 = 81$

    $4^3 = 4 \times 4 \times 4 \quad = 64$

    $3^4 > 4^3$

3. Express 243 as a power of 10.

    $243 = 200 + 40 + 3$ ←— write in expanded form first

    $= 2 \times 100 + 4 \times 10 + 3$

    $= 2 \times 10^2 + 4 \times 10^1 + 3 \times 10^0$

H INTS:

- Express 10, 100, 1000, or ... as a power of 10, the exponent is equal to the number of zeros after 1.

    e.g. $1000 = 10^3$; $10\ 000 = 10^4$

- Any power with an exponent of 0 has the value of 1.

    e.g. $6^0 = 1$; $109^0 = 1$

Complete the table.

	Product	Power	Base	Exponent	Value
①	$4 \times 4 \times 4 \times 4 \times 4$				
②	$3 \times 3 \times 3$				
③	$6 \times 6 \times 6$				
④		2^4			
⑤		7^2			
⑥			3	6	
⑦			8	4	

  ISBN: 978-1-897164-21-1

Find the values.

⑧ $10^4 =$ _____

⑨ $9^3 =$ _____

⑩ $45^0 =$ _____

⑪ $7^3 =$ _____

⑫ $16^1 =$ _____

⑬ $279^1 =$ _____

⑭ $2^5 =$ _____

⑮ $8^3 =$ _____

⑯ $533^0 =$ _____

Complete each of the following statements using '>', '<', or '='.

⑰ 4^3 ☐ 4^5

⑱ 2^4 ☐ 2^6

⑲ 3^1 ☐ 3^0

⑳ 1^8 ☐ 1^6

㉑ 8^3 ☐ 9^3

㉒ 7^3 ☐ 3^7

Write each group of powers in order from least to greatest.

㉓ 2^3 2^2 2^1 2^4

_____

㉔ 1^5 2^7 3^2 5^7

_____

Write each number as a power of 10 or find the values.

㉕ 2956 = ___ $\times 10^3 +$ ___ $\times 10^2 +$ ___ $\times 10^1 +$ ___ $\times 10^0$

㉖ $12\ 468$ = $1 \times$ ___ $+ 2 \times$ ___ $+ 4 \times$ ___ $+ 6 \times$ ___ $+ 8 \times$ ___

㉗ 7361 = _____

㉘ $9 \times 10^5 + 4 \times 10^4 + 6 \times 10^3 + 9 \times 10^2 + 1 \times 10^1 + 2 \times 10^0 =$ _____

㉙ $8 \times 10^5 + 3 \times 10^3 + 6 \times 10^0 =$ _____

Write the expanded form of each number. Then write it as a power of 10.

㉚ 875 = _____ + _____ + _____

=

㉛ 7180 =

㉜ 1024 =

㉝ $59\ 003$ =

ISBN: 978-1-897164-21-1

2 Squares and Square Roots

1. Complete the factor tree.

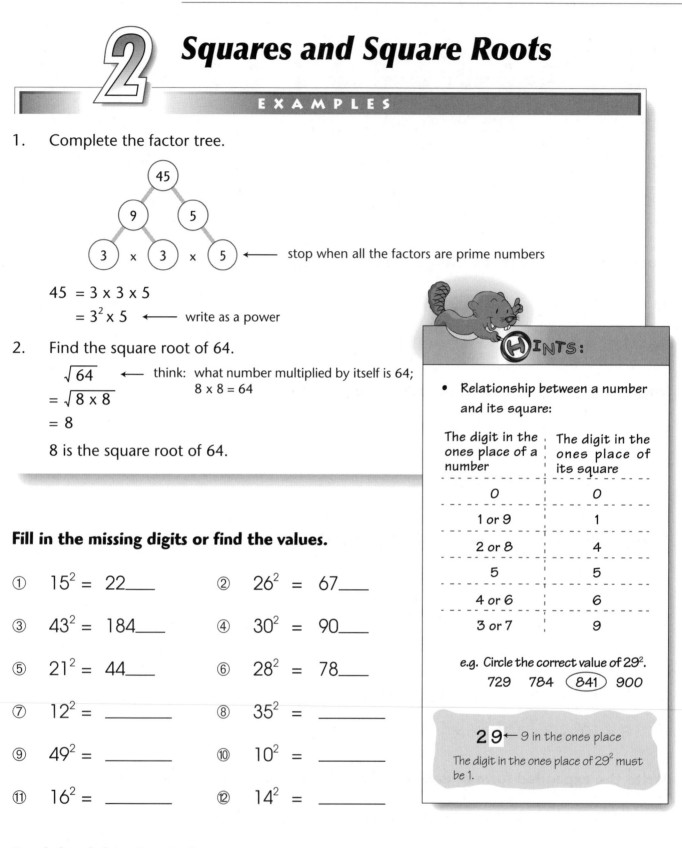

```
        45
       /  \
      9    5
     / \    \
    3 x 3 x 5  ←——— stop when all the factors are prime numbers
```

45 = 3 x 3 x 5
  = 3^2 x 5 ←——— write as a power

2. Find the square root of 64.

$\sqrt{64}$ ←——— think: what number multiplied by itself is 64;
8 x 8 = 64

$= \sqrt{8 \times 8}$

$= 8$

8 is the square root of 64.

HINTS:

- Relationship between a number and its square:

The digit in the ones place of a number	The digit in the ones place of its square
0	0
1 or 9	1
2 or 8	4
5	5
4 or 6	6
3 or 7	9

e.g. Circle the correct value of 29^2.
729   784   (841)   900

2 9 ←— 9 in the ones place

The digit in the ones place of 29^2 must be 1.

Fill in the missing digits or find the values.

① $15^2 = 22\_\_\_$

② $26^2 = 67\_\_\_$

③ $43^2 = 184\_\_\_$

④ $30^2 = 90\_\_\_$

⑤ $21^2 = 44\_\_\_$

⑥ $28^2 = 78\_\_\_$

⑦ $12^2 = _____$

⑧ $35^2 = _____$

⑨ $49^2 = _____$

⑩ $10^2 = _____$

⑪ $16^2 = _____$

⑫ $14^2 = _____$

Put '=' or '≠' in the circles.

⑬ $\sqrt{170}$ ◯ 13

⑭ $\sqrt{196}$ ◯ 14

⑮ $\sqrt{44}$ ◯ 22

⑯ 9 ◯ $\sqrt{99}$

⑰ 28 ◯ $\sqrt{784}$

⑱ $\sqrt{400}$ ◯ 20

ISBN: 978-1-897164-21-1

Write each number in an exponent form by using factor tree.

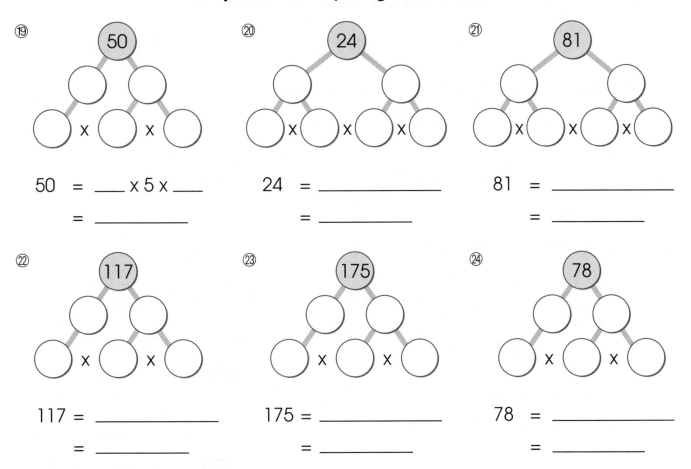

19)

50 = ___ x 5 x ___

= _____

20)

24 = _____

= _____

21)

81 = _____

= _____

22)

117 = _____

= _____

23)

175 = _____

= _____

24)

78 = _____

= _____

Find the square roots. Use arrows to locate the numbers on the number line. Then fill in the blanks with the help of the number line.

25) $\sqrt{4}$ = $\sqrt{\rule{0.5cm}{0.15mm} \times \rule{0.5cm}{0.15mm}}$ = _____

26) $\sqrt{16}$ = _____ = _____

27) $\sqrt{100}$ = _____ = _____

28) $\sqrt{121}$ = _____ = _____

29) $\sqrt{169}$ = _____ = _____

30) $\sqrt{196}$ = _____ = _____

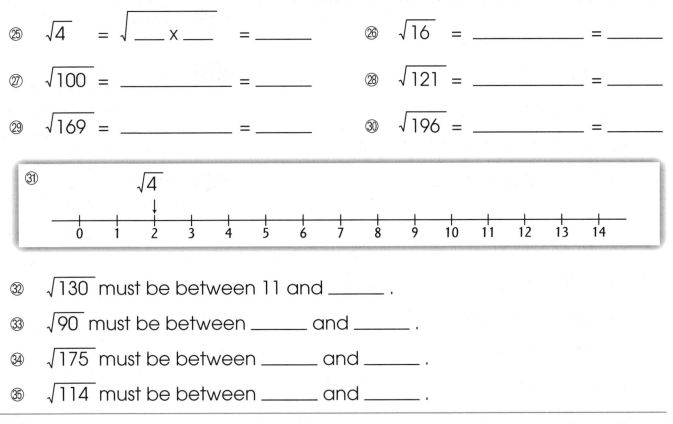

31)

$\sqrt{4}$

0 1 2 3 4 5 6 7 8 9 10 11 12 13 14

32) $\sqrt{130}$ must be between 11 and _____ .

33) $\sqrt{90}$ must be between _____ and _____ .

34) $\sqrt{175}$ must be between _____ and _____ .

35) $\sqrt{114}$ must be between _____ and _____ .

ISBN: 978-1-897164-21-1

Multiples and Factors

1. Find the greatest common factor (G.C.F.) of 6, 12, and 18.

2	6	12	18	← 2 is their common factor
3	3	6	9	← write down the quotients; divide them by their common factor 3
	1	2	3	← stop here because the numbers have no common factors

   G.C.F. of 6, 12, and 18 = ② x ③ = 6

2. Find the least common multiple (L.C.M.) of 4, 12, and 18.

2	4	12	18	← 2 is their G.C.F.
2	2	6	9	← 2 is the common factor of 2 and 6
3	1	3	9	← 3 is the common factor of 3 and 9
	1	1	3	← stop here

   L.C.M. of 4, 12, and 18

   = ② x ② x ③ x ① x ① x ③

   = 36

Find the G.C.F. or L.C.M. of each group of numbers.

① G.C.F. of 12 and 20

   Factors of 12: _____

   Factors of 20: _____

   Common factors of 12 and 20: _____

   G.C.F. of 12 and 20: _____

· ·

② L.C.M. of 6 and 15

   Multiples of 6 (up to 100): _____

   Multiples of 15 (up to 100): _____

   Common multiples of 6 and 15 (up to 100): _____

   L.C.M. of 6 and 15: _____

HINTS:

- Steps for finding **G.C.F.** :

  1st Keep dividing the numbers by their common factors until the numbers have no common factors.

  2nd Multiply all the divisors to get the G.C.F.

- Steps for finding **L.C.M.** :

  1st Divide the numbers by their G.C.F.

  2nd Keep dividing the numbers by the common factors of any two or more of the numbers until the numbers have no common factors.

  3rd Multiply all the divisors and the last row of quotients to get the L.C.M.

 ISBN: 978-1-897164-21-1

Complete the following to find the G.C.F.

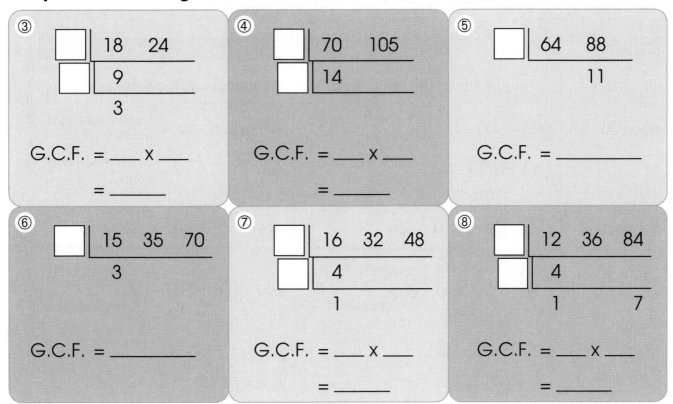

③
```
  ☐ | 18   24
  ☐ | 9
      3
```
G.C.F. = ___ x ___

= _____

④
```
  ☐ | 70   105
  ☐ | 14
```
G.C.F. = ___ x ___

= _____

⑤
```
  ☐ | 64   88
            11
```
G.C.F. = _____

⑥
```
  ☐ | 15   35   70
      3
```
G.C.F. = _____

⑦
```
  ☐ | 16   32   48
  ☐ | 4
      1
```
G.C.F. = ___ x ___

= _____

⑧
```
  ☐ | 12   36   84
  ☐ | 4
      1         7
```
G.C.F. = ___ x ___

= _____

Complete the following to find the L.C.M.

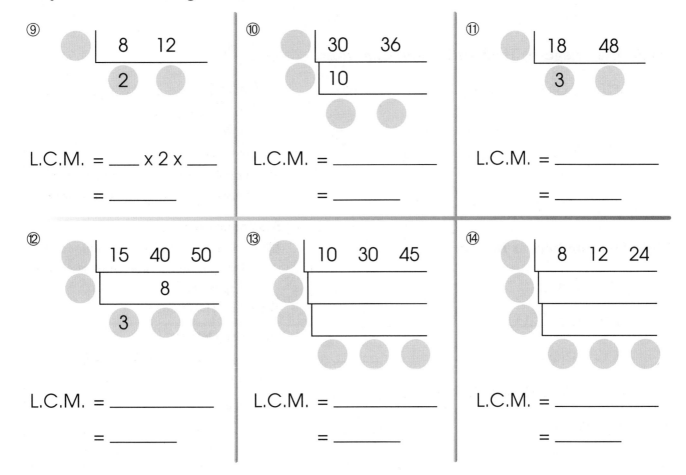

⑨
```
  ○ | 8   12
      2   ○
```
L.C.M. = ___ x 2 x ___

= _____

⑩
```
  ○ | 30   36
  ○ | 10
      ○    ○
```
L.C.M. = _____

= _____

⑪
```
  ○ | 18   48
      3   ○
```
L.C.M. = _____

= _____

⑫
```
  ○ | 15   40   50
  ○ | 8
      3   ○   ○
```
L.C.M. = _____

= _____

⑬
```
  ○ | 10   30   45
  ○ |
  ○ |
      ○   ○   ○
```
L.C.M. = _____

= _____

⑭
```
  ○ | 8   12   24
  ○ |
  ○ |
      ○   ○   ○
```
L.C.M. = _____

= _____

ISBN: 978-1-897164-21-1

4 Integers

	Integers with the same signs	Integers with different signs
Addition	$(+5) + (+7) = +12$ ← count 7 forward from +5 $(-5) + (-7) = -12$ ← count 7 backward from -5	$(+5) + (-7) = -2$ ← count 7 backward from +5 $(-5) + (+7) = +2$ ← count 7 forward from -5
Subtraction	$(+5) - (+7) = (+5) + (-7)$ — change '−' to '+' / opposite of +7 $= -2$ $(-5) - (-7) = (-5) + (+7)$ — change '−' to '+' / opposite of -7 $= +2$	$(+5) - (-7) = (+5) + (+7)$ — change '−' to '+' / opposite of -7 $= +12$ $(-5) - (+7) = (-5) + (-7)$ — change '−' to '+' / opposite of +7 $= -12$
Multiplication	$(+5) \times (+7) = +35$ $(-5) \times (-7) = +35$	$(+5) \times (-7) = -35$ $(-5) \times (+7) = -35$
Division	$(+35) \div (+5) = +7$ $(-35) \div (-5) = +7$	$(+35) \div (-5) = -7$ $(-35) \div (+5) = -7$

Circle the integers. Then write their opposites.

① -15   0.04 $\sqrt{5}$ 7   -6

___   ___   ___   ___   ___

$4\frac{1}{2}$ 8.1   120   -3.3   29

___   ___   ___   ___   ___

Write the numbers as integers.

② gain 45 kg _____

③ withdraw $200 _____

④ 12°C below 0°C _____

⑤ 140 m above sea level _____

⑥ deposit $35 _____

⑦ lose 2 kg _____

HINTS:

- Adding integers:
  Add a positive integer → Count forward
  Add a negative integer → Count backward

- Subtracting integers:
  1st Change '−' to '+' and replace the integer after the sign with its opposite.
  2nd Add the integers.

- Multiplying integers:
  $(+) \times (+) = (+)$ $(+) \times (-) = (-)$
  $(-) \times (-) = (+)$ $(-) \times (+) = (-)$

- Dividing integers:
  $(+) \div (+) = (+)$ $(+) \div (-) = (-)$
  $(-) \div (-) = (+)$ $(-) \div (+) = (-)$

Write the integers.

⑧ An integer greater than -2 _____

⑨ An integer less than -5 _____

⑩ An integer greater than -3 but less than +2 _____

⑪ An integer between -4 and +3 _____

⑫ An integer not between 8 and -4 _____

⑬ The opposite of +6 _____

⑭ The opposite of -5 _____

Write each group of integers in order from least to greatest.

⑮ -14, -38, 19, -4, 6

_____

⑯ 15 m, 16 m, -4 m, -3 m, 2 m

_____

⑰ 23, -3, 11, -15, 0

_____

⑱ -5°C, 9°C, 32°C, -7°C, 0°C

_____

Check ✔ the correct letter to match each diagram.

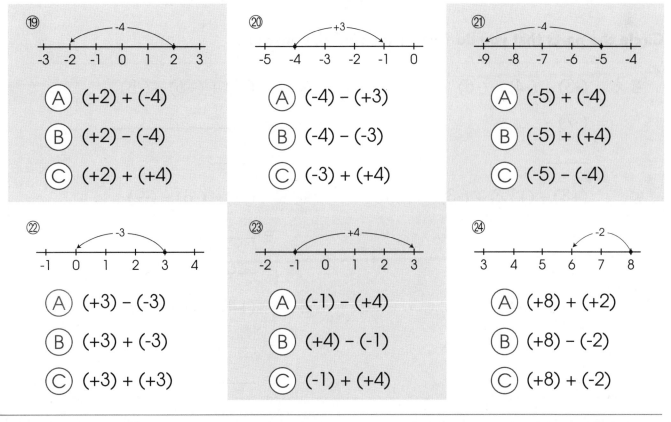

⑲
-3 -2 -1 0 1 2 3   (-4)

Ⓐ (+2) + (-4)

Ⓑ (+2) - (-4)

Ⓒ (+2) + (+4)

⑳
-5 -4 -3 -2 -1 0   (+3)

Ⓐ (-4) - (+3)

Ⓑ (-4) - (-3)

Ⓒ (-3) + (+4)

㉑
-9 -8 -7 -6 -5 -4   (-4)

Ⓐ (-5) + (-4)

Ⓑ (-5) + (+4)

Ⓒ (-5) - (-4)

㉒
-1 0 1 2 3 4   (-3)

Ⓐ (+3) - (-3)

Ⓑ (+3) + (-3)

Ⓒ (+3) + (+3)

㉓
-2 -1 0 1 2 3   (+4)

Ⓐ (-1) - (+4)

Ⓑ (+4) - (-1)

Ⓒ (-1) + (+4)

㉔
3 4 5 6 7 8   (-2)

Ⓐ (+8) + (+2)

Ⓑ (+8) - (-2)

Ⓒ (+8) + (-2)

ISBN: 978-1-897164-21-1

Do the addition with the help of the number line.

```
+----+----+----+----+----+----+----+----+----+----+----+----+----+----+----+----+----+----+----+----+
-10  -9   -8   -7   -6   -5   -4   -3   -2   -1   0    1    2    3    4    5    6    7    8    9   10
```

㉕ $(+3) + (+6) =$ _____   ㉖ $(-4) + (-1) =$ _____

㉗ $(-8) + (-1) =$ _____   ㉘ $(-6) + (-4) =$ _____

㉙ $(-4) + (-4) =$ _____   ㉚ $(+7) + (-3) =$ _____

㉛ $(-6) + (+7) =$ _____   ㉜ $(+5) + (-9) =$ _____

㉝ $(+3) + (-1) =$ _____   ㉞ $(-10) + (+3) =$ _____

Do the subtraction.

㉟ $(+14) - (+7)$ ㊱ $(-5) - (-6)$ ㊲ $(+7) - (-1)$

= $(+14) + (-7)$ = $(-5) + (+6)$ = _____

= _____   = _____   = _____

㊳ $(-7) - (+11)$ ㊴ $(-13) - (-2)$ ㊵ $(-9) - (+7)$

= _____   = _____   = _____

= _____   = _____   = _____

Circle the parts that you do first. Then find the answers.

㊶ $(-3) + (-4) + (-5)$ ㊷ $(-1) + (+6) - (-8)$

= $(-7) +$ _____   = _____

= _____   = _____

㊸ $(+8) - (+12) + (-3)$ ㊹ $(-7) - (-8) - (-9)$

= _____   = _____

= _____   = _____

㊺ $(-10) + (-6) - (+4)$ ㊻ $(+4) - (-8) + (-3)$

= _____   = _____

= _____   = _____

Do the multiplication.

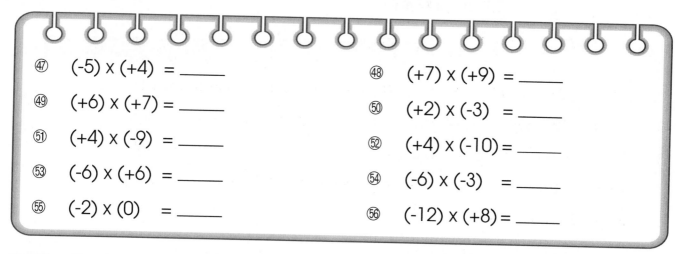

(47) $(-5) \times (+4) =$ ____

(48) $(+7) \times (+9) =$ ____

(49) $(+6) \times (+7) =$ ____

(50) $(+2) \times (-3) =$ ____

(51) $(+4) \times (-9) =$ ____

(52) $(+4) \times (-10) =$ ____

(53) $(-6) \times (+6) =$ ____

(54) $(-6) \times (-3) =$ ____

(55) $(-2) \times (0) =$ ____

(56) $(-12) \times (+8) =$ ____

Do the division.

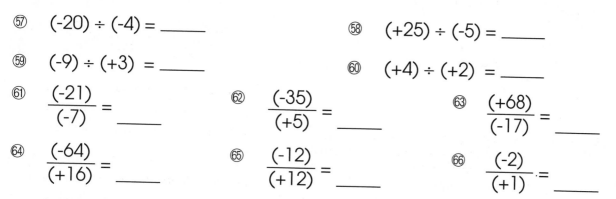

(57) $(-20) \div (-4) =$ ____

(58) $(+25) \div (-5) =$ ____

(59) $(-9) \div (+3) =$ ____

(60) $(+4) \div (+2) =$ ____

(61) $\frac{(-21)}{(-7)} =$ ____

(62) $\frac{(-35)}{(+5)} =$ ____

(63) $\frac{(+68)}{(-17)} =$ ____

(64) $\frac{(-64)}{(+16)} =$ ____

(65) $\frac{(-12)}{(+12)} =$ ____

(66) $\frac{(-2)}{(+1)} =$ ____

Circle the parts that you do first. Then find the answers.

(67) $(-5) - (-2) \times (-4)$

$=$ ____ $- (+8)$

$=$ ____

(68) $(-10) + (-6) \div (+2)$

$=$

(69) $(-9) \div (-3) + (-9)$

$=$

(70) $(-14) \div (-2) + (-6)$

$=$

(71) $(-4) + (-2) \times (-4)$

$=$

(72) $(-3) \times (-6) \div (-2)$

$=$

(73) $(-25) + (-9) \times (-2)$

$=$

(74) $(-2) - (-12) \div (-6)$

$=$

(75) $(-8) \div (-4) \times (-10)$

$=$

ISBN: 978-1-897164-21-1

Decimals

1. 64.3 x 0.01 = 0.643   ← move the decimal point 2 places to the left

2. 5.9 x 1000 = 5900.   ← move the decimal point 3 places to the right

3. 1.256 ÷ 0.001 = 1256.   ← move the decimal point 3 places to the right

4. 25.8 ÷ 100 = 0.258   ← move the decimal point 2 places to the left

5. 6.89 x 3.2 = 22.048

$$
\begin{array}{r}
6.89 \quad \leftarrow \text{2 decimal places} \\
\times \quad 3.2 \quad \leftarrow \text{1 decimal place} \\
\hline
1\,378 \\
2\,0\,6\,7\,0 \\
\hline
2\,2.0\,4\,8 \quad \leftarrow \text{3 decimal places}
\end{array}
$$

6. 0.133 ÷ 0.7
    = 1.33 ÷ 7   ← change the divisor to a whole number first; move the decimal points in both numbers 1 place to the right
    = 0.19

HINTS:

- The number of decimal places in the product is equal to the sum of the decimal places in the two numbers.

- To divide a decimal by another decimal, change the divisor to a whole number first. Then do the division.

Write True (T) or False (F) in the circles.

① 8.04 = 8.4   ◯

② 6.30 = 6.3   ◯

③ 4.804 > 4.084   ◯

④ 7.53 < 7.053   ◯

⑤ 2.584 = 2 + 0.05 + 0.008 + 0.0004   ◯

⑥ 17.189 = 10 + 7 + 0.1 + 0.08 + 0.009   ◯

Put the decimal points in the correct places. Add zero(s) if necessary.

⑦ 54 x 0.01   = _____ 5 4 _____

⑧ 12 ÷ 0.1   = _____ 1 2 _____

⑨ 2.075 x 100 = _____ 2 0 7 5 _____

⑩ 146 ÷ 0.01 = _____ 1 4 6 _____

⑪ 3.005 ÷ 0.01 = _____ 3 0 0 5 _____

⑫ 0.068 x 100 = _____ 6 8 _____

⑬ 2.889 x 0.1 = _____ 2 8 8 9 _____

⑭ 11.64 x 10 = _____ 1 1 6 4 _____

ISBN: 978-1-897164-21-1

Find the values.

⑮ 5.867 x 0.001 =	⑯ 0.012 x 10 =
⑰ 0.219 x 0.01 =	⑱ 548 x 0.001 =
⑲ 0.0084 ÷ 0.01 =	⑳ 2.413 ÷ 10 =
㉑ 19.01 ÷ 0.001 =	㉒ 0.315 ÷ 100 =

Add or subtract.

㉓
```
    70.86
  +  7.07
```

㉔
```
    98.04
  -  6.10
```

㉕
```
    100
  -   6.23
```

㉖ 9.02 + 92   = _____

㉗ 6.71 + 68.9 = _____

㉘ 603 + 7.28   = _____

㉙ 400 – 5.87   = _____

㉚ 6.96 – 4.106 = _____

㉛ 701.4 – 2.13 = _____

Circle the parts that you do first. Then find the answers.

㉜ (67.08 + 9.2) – 6.72

= _____ – 6.72

= _____

㉝ 21.42 – 16.9 + 7.65

= _____

= _____

㉞ 722.9 – 69 + 409.12

= _____

= _____

㉟ 26.99 – 18.12 – 8.86

= _____

= _____

㊱ 56.45 + 47.58 – 20.98

= _____

= _____

㊲ 600 – 124.8 – 469.91

= _____

= _____

Place the decimal point in each product and find the answer. Add zero(s) if necessary.

㊳
$$18.2$$
$$\times \quad 0.3$$
$$\overline{\quad 5\,4\,6}$$

㊴
$$11.93$$
$$\times \quad 0.04$$
$$\overline{\quad 4\,7\,7\,2}$$

㊵
$$7.11$$
$$\times \quad 0.006$$
$$\overline{\quad 4\,2\,6\,6}$$

㊶
$$0.096$$
$$\times \qquad 9$$
$$\overline{\qquad}$$

㊷
$$3.08$$
$$\times \quad 0.26$$
$$\overline{\quad 1\,8\,4\,8}$$
$$\underline{\quad 6\,1\,6\,0}$$

㊸
$$0.253$$
$$\times \quad 0.14$$
$$\overline{\qquad}$$

Multiply.

㊹ $68.02 \times 0.4 =$ _____

㊺ $2.43 \times 1.6 \ =$ _____

㊻ $104.1 \times 3.2 =$ _____

㊼ $1.92 \times 0.35 =$ _____

㊽ $5.76 \times 0.19 =$ _____

㊾ $3.43 \times 26 \ =$ _____

Solve the problems.

㊿ Each bag of candies costs $7.99. If Mary buys 3 bags of candies, how much does she pay?

_____

51 Each box of chocolates costs $8.99. If Mary's sister, Judy, buys 2 boxes of chocolates, how much does she pay?

_____

52 If Mary's mother pays for the things that Mary and Judy want to buy, how much does she need to pay?

_____

53 If Mary's mother pays $50 to the cashier, what is her change?

_____

   ISBN: 978-1-897164-21-1

Rewrite the division sentences. Then use long division to find the answers.

(54) $1.08 \div 0.6$

$= 10.8 \div 6$

$=$ ______

$6\overline{)10.8}$

(55) $0.768 \div 0.08$

$=$ ______

$=$ ______

(56) $625 \div 2.5$

$=$ ______

$=$ ______

Divide.

(57) $2.24 \div 0.07 =$ ______

(58) $0.1875 \div 0.75 =$ ______

(59) $5390 \div 1.1 =$ ______

(60) $40.5 \div 0.9 =$ ______

(61) $6.24 \div 0.4 =$ ______

(62) $4.25 \div 1.7 =$ ______

Circle the parts that you do first. Then find the answers.

(63) $16.39 - (6.2 \times 0.7)$

$= 16.39 -$ ______

$=$ ______

(64) $4.2 + (5 \times 1.6)$

$=$ ______

$=$ ______

(65) $67.5 \div 0.05 \div 0.2$

$=$ ______

$=$ ______

(66) $53.9 - 12.8 \div 3.2$

$=$ ______

$=$ ______

Find the average of each group of numbers.

(67) 8.4, 3.7, 6.38

Average: ______

(68) 4.27, 13.28, 9.15, 5.62

Average: ______

(69) 4.91, 1.35, 6.5, 3.7, 5.49

Average: ______

ISBN: 978-1-897164-21-1

Fractions

1. $8\frac{4}{5} - 3\frac{1}{8} \times 1\frac{1}{5}$ ← do multiplication first

    $= 8\frac{4}{5} - \frac{25\,^5}{8\,_4} \times \frac{6\,^3}{5\,_1}$ ← change the mixed numbers to improper fractions; simplify by dividing the numerators and denominators by their common factors

    $= 8\frac{4}{5} - \frac{15}{4}$ ← multiply the numerators and the denominators to get $\frac{15}{4}$

    $= 8\frac{4}{5} - 3\frac{3}{4}$ ← change back to a mixed number

    $= 8\frac{16}{20} - 3\frac{15}{20}$ ← write equivalent fractions with common denominator 20

    $= 5\frac{1}{20}$ ← subtract the whole numbers and fractions separately

2. $4\frac{5}{6} + 1\frac{5}{9} \div 2\frac{1}{3}$ ← do division first

    $= 4\frac{5}{6} + \frac{14}{9} \div \frac{7}{3}$ ← change the mixed numbers to improper fractions

    $= 4\frac{5}{6} + \frac{14\,^2}{9\,_3} \times \frac{3\,^1}{7\,_1}$ ← change '÷' to 'x' and the divisor to its reciprocal; then simplify

    $= 4\frac{5}{6} + \frac{2}{3}$ ← multiply the numerators and the denominators to get $\frac{2}{3}$

    $= 4\frac{5}{6} + \frac{4}{6}$ ← write equivalent fractions with common denominator 6

    $= 4\frac{9}{6}$ ← add the numerators

    $= 5\frac{1}{2}$ ← reduce the answer to the lowest terms

HINTS:

- Addition and Subtraction of Fractions:

  1st Write the fractions with the same denominators.

  2nd Add or subtract the fractions.

  3rd Reduce the answer to lowest terms.

- Multiplication of Fractions:

  1st Change the mixed numbers to improper fractions.

  2nd Simplify by the common factors.

  3rd Multiply.

  4th Reduce the answer to lowest terms.

- Division of Fractions:

  1st Change the mixed numbers to improper fractions.

  2nd Change '÷' to 'x' and the divisor to its reciprocal.

  3rd Multiply.

  4th Reduce the answer to lowest terms.

Write each improper fraction as a mixed number in lowest terms.

① $\frac{20}{18} =$ _____

② $\frac{36}{20} =$ _____

③ $\frac{105}{63} =$ _____

④ $\frac{110}{72} =$ _____

⑤ $\frac{68}{48} =$ _____

⑥ $\frac{45}{15} =$ _____

⑦ $\frac{136}{4} =$ _____

⑧ $\frac{183}{9} =$ _____

ISBN: 978-1-897164-21-1

Write each group of fractions from greatest to least using '>'.

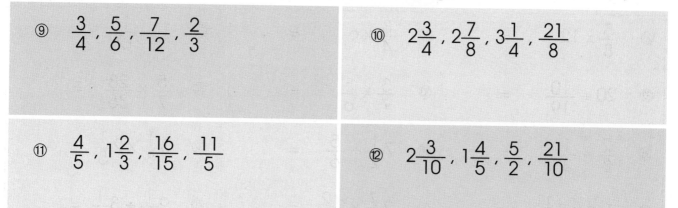

⑨ $\dfrac{3}{4}$, $\dfrac{5}{6}$, $\dfrac{7}{12}$, $\dfrac{2}{3}$

⑩ $2\dfrac{3}{4}$, $2\dfrac{7}{8}$, $3\dfrac{1}{4}$, $\dfrac{21}{8}$

⑪ $\dfrac{4}{5}$, $1\dfrac{2}{3}$, $\dfrac{16}{15}$, $\dfrac{11}{5}$

⑫ $2\dfrac{3}{10}$, $1\dfrac{4}{5}$, $\dfrac{5}{2}$, $\dfrac{21}{10}$

Find the sums or differences. Reduce the answers to lowest terms.

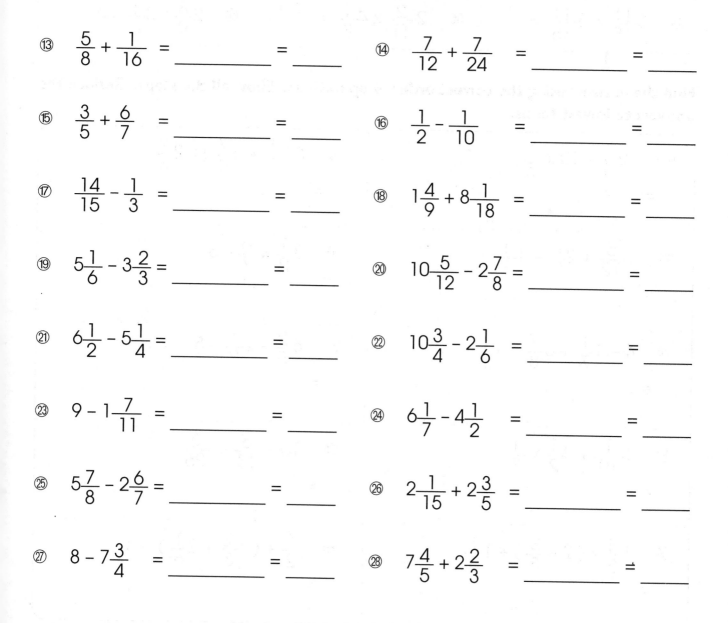

⑬ $\dfrac{5}{8} + \dfrac{1}{16}$ = _____ = ____

⑭ $\dfrac{7}{12} + \dfrac{7}{24}$ = _____ = ____

⑮ $\dfrac{3}{5} + \dfrac{6}{7}$ = _____ = ____

⑯ $\dfrac{1}{2} - \dfrac{1}{10}$ = _____ = ____

⑰ $\dfrac{14}{15} - \dfrac{1}{3}$ = _____ = ____

⑱ $1\dfrac{4}{9} + 8\dfrac{1}{18}$ = _____ = ____

⑲ $5\dfrac{1}{6} - 3\dfrac{2}{3}$ = _____ = ____

⑳ $10\dfrac{5}{12} - 2\dfrac{7}{8}$ = _____ = ____

㉑ $6\dfrac{1}{2} - 5\dfrac{1}{4}$ = _____ = ____

㉒ $10\dfrac{3}{4} - 2\dfrac{1}{6}$ = _____ = ____

㉓ $9 - 1\dfrac{7}{11}$ = _____ = ____

㉔ $6\dfrac{1}{7} - 4\dfrac{1}{2}$ = _____ = ____

㉕ $5\dfrac{7}{8} - 2\dfrac{6}{7}$ = _____ = ____

㉖ $2\dfrac{1}{15} + 2\dfrac{3}{5}$ = _____ = ____

㉗ $8 - 7\dfrac{3}{4}$ = _____ = ____

㉘ $7\dfrac{4}{5} + 2\dfrac{2}{3}$ = _____ = ____

Multiply or divide. Reduce the answers to lowest terms.

㉙ $\frac{5}{8} \times 12$ =	㉚ $\frac{3}{4} \times 6$ =	㉛ $1 \div \frac{1}{12}$ =
㉜ $20 \div \frac{10}{19}$ =	㉝ $\frac{3}{7} \times \frac{5}{6}$ =	㉞ $\frac{5}{7} \times \frac{28}{35}$ =
㉟ $\frac{2}{3} \div \frac{1}{3}$ =	㊱ $7\frac{1}{2} \div \frac{5}{6}$ =	㊲ $3\frac{1}{8} \times 2\frac{1}{5}$ =
㊳ $1\frac{1}{5} \times 3\frac{1}{3}$ =	㊴ $2\frac{7}{8} \times 2\frac{2}{5}$ =	㊵ $9\frac{1}{3} \div 3\frac{1}{9}$ =
㊶ $6\frac{1}{7} \div 1\frac{11}{14}$ =	㊷ $1\frac{7}{18} \div 2\frac{2}{9}$ =	㊸ $4\frac{3}{4} \div 5\frac{7}{8}$ =
㊹ $3\frac{11}{15} \div 3\frac{11}{15}$ =	㊺ $2\frac{2}{11} \times 4\frac{1}{8}$ =	㊻ $2\frac{8}{9} \times 3\frac{1}{6}$ =

Find the answers using the correct order of operations. Show all the steps. Reduce the answers to lowest terms.

㊼ $12\frac{2}{5} - 10 \times \frac{1}{5}$

=

㊽ $(1\frac{1}{2} + 5\frac{7}{8}) \times 2\frac{2}{3}$

=

㊾ $1\frac{5}{12} + 2\frac{4}{9} \div 1\frac{1}{3}$

=

㊿ $3\frac{1}{2} \times \frac{6}{7} - 3$

=

(51) $4 \div 1\frac{1}{7} + 6\frac{7}{8}$

=

(52) $5\frac{2}{5} - 2\frac{1}{2} \div 5$

=

(53) $1\frac{1}{10} + 4\frac{1}{2} \times \frac{1}{3}$

=

(54) $10 - \frac{6}{13} \div \frac{3}{26}$

=

(55) $1\frac{4}{5} \times (2 - \frac{3}{5}) \div 1\frac{1}{5}$

=

(56) $\frac{1}{2} + (1\frac{2}{3} + 2\frac{1}{6}) \div 3$

=

ISBN: 978-1-897164-21-1

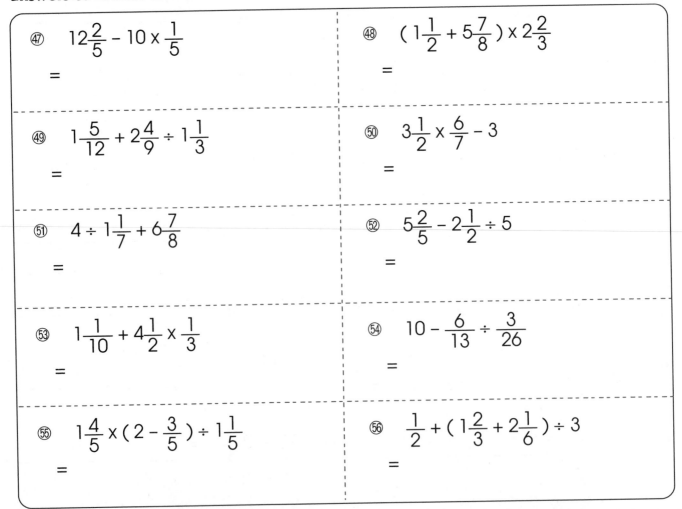

Look at the recipe and solve the problems.

㊄

Egg Salad Sandwich (4 servings)

* Preparation Time: $\frac{1}{3}$ hour
* Cooking Time: $\frac{1}{4}$ hour

$\frac{2}{3}$ carton of eggs

$\frac{3}{5}$ of a red onion

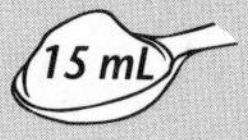

$1\frac{1}{2}$ tablespoons of mayonnaise

$\frac{2}{5}$ bag of bread

Time in min:

a.
_____ X _____ = _____ ; _____ min

b.
_____ X _____ = _____ ; _____ min

c. No. of eggs:
_____ X _____ = _____

d. Weight of onion:
_____ X _____ = _____ (g)

e. Amount of mayonnaise:
_____ X _____ = _____ (mL)

f. Slices of bread:
_____ X _____ = _____

㊄ If $\frac{1}{4}$ of an onion weighs 32 g, how heavy is the onion?

_____ = _____   _____ g

㊄ If there are 6 eggs in $\frac{3}{4}$ carton, how many eggs does a carton hold?

_____ = _____   _____ eggs

㊅ If Sara doubles the preparation and cooking times shown above to make 4 servings of sandwich, how many hours will Sara take?

_____ = _____   _____ h

㊅ If Sara wants to change the recipe from 4 servings to 3 servings, help her find out the correct amount of each ingredient.

Egg Salad Sandwich * 3 servings

No. of eggs : _____

Red onion : _____ of an onion

Amount of mayonnaise : _____ tablespoons

Slices of bread : _____

$\times \frac{3}{4} =$

$\frac{3}{5} \times$ _____ =

$1\frac{1}{2} \times$ _____ =

$\times \frac{3}{4} =$

ISBN: 978-1-897164-21-1

7 Ratios and Rates

1. Uncle Tim drives 150 km in 2 hours and Aunt Sue drives 320 km in 4 hours. Who drives at a higher speed?

   Uncle Tim's speed: $150 \div 2 = 75$ (km/h) ⟵ travelling 75 km in 1 h
   Aunt Sue's speed: $320 \div 4 = 80$ (km/h) ⟵ travelling 80 km in 1 h
   Aunt Sue drives at a higher speed.

2. What is the ratio of number of blocks in the box to that in the bag?

   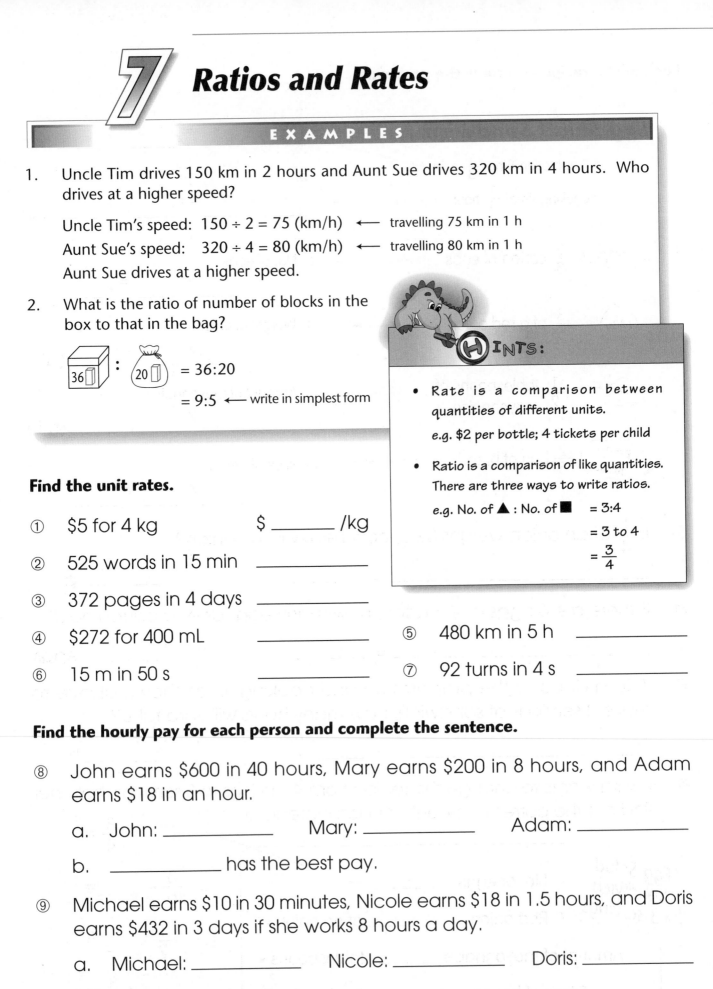

   $36 : 20 = 36:20$

   $= 9:5$ ⟵ write in simplest form

HINTS:

- Rate is a comparison between quantities of different units.
  e.g. $2 per bottle; 4 tickets per child

- Ratio is a comparison of like quantities. There are three ways to write ratios.
  e.g. No. of ▲ : No. of ■ = 3:4
  = 3 to 4
  = $\frac{3}{4}$

Find the unit rates.

① $5 for 4 kg $ _____ /kg

② 525 words in 15 min   _____

③ 372 pages in 4 days   _____

④ $272 for 400 mL   _____   ⑤ 480 km in 5 h   _____

⑥ 15 m in 50 s   _____   ⑦ 92 turns in 4 s   _____

Find the hourly pay for each person and complete the sentence.

⑧ John earns $600 in 40 hours, Mary earns $200 in 8 hours, and Adam earns $18 in an hour.

   a. John: _____   Mary: _____   Adam: _____

   b. _____ has the best pay.

⑨ Michael earns $10 in 30 minutes, Nicole earns $18 in 1.5 hours, and Doris earns $432 in 3 days if she works 8 hours a day.

   a. Michael: _____   Nicole: _____   Doris: _____

   b. _____ has the best pay.

ISBN: 978-1-897164-21-1

Find the unit rates and circle the better buy.

⑩ $62.19 for 3 kg ➡ $ _____ /kg

$119.70 for 5 kg ➡ $ _____ /kg

⑪ $77.31 for 9 bags ➡ $ _____ /bags

$20.16 for 3 bags ➡ $ _____ /bags

⑫ $2.76 for 2.4 m ➡ $ _____ /m

$2.31 for 150 cm ➡ $ _____ /m

⑬ $15 for 0.5 L ➡ $ _____ /mL

$12 for 240 mL ➡ $ _____ /mL

Write each ratio in simplest form in 2 different ways.

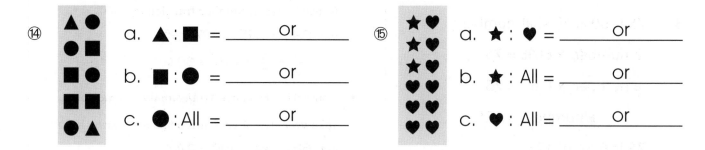

⑭   a. ▲ : ■ = _____ or _____

     b. ■ : ● = _____ or _____

     c. ● : All = _____ or _____

⑮   a. ★ : ♥ = _____ or _____

     b. ★ : All = _____ or _____

     c. ♥ : All = _____ or _____

Write each ratio in simplest form to complete the table. Then answer the questions.

Hockey Team	Win	Loss	Tie	Total No. of Games	Wins to Games	Losses to Games	Ties to Games
⑯ Terminators	11	4	2				
⑰ Bears	12	3	3				
⑱ Raiders	10	2	3				
⑲ Wizards	15	1	1				

⑳ Which two teams have the same ratios of 'Wins' to 'Games'?

_____

㉑ If the Wizards played 1 more game and lost, what would be the new ratio of

     a.   'Wins' to 'Games'?   _____

     b.   'Losses' to 'Games'?   _____

ISBN: 978-1-897164-21-1

Percents

1. What percent is 17 out of 40?

$\dfrac{17}{40}_2 \times \overset{5}{\cancel{100}}\% = 42.5\%$

17 out of 40 is 42.5%.

2. What is 16% of 25?

$25 \times 16\% = 25 \times 0.16 = 4$

16% of 25 is 4.

3. 75 is 60% of what number?

a number $\times 60\% = 75$

a number $\times 0.6 = 75$

a number $= 125$

75 is 60% of 125.

HINTS:

- Converting Decimals or Fractions to Percents:

  Multiply the decimal or fraction by 100%.

  e.g. $0.45 = 0.45 \times 100\% = 45\%$

  $\dfrac{3}{5} = \dfrac{3}{5}_1 \times \overset{20}{\cancel{100}}\% = 60\%$

- Converting Percents to Decimals or Fractions:

  Take away the % and divide the number by 100.

  e.g. $68\% = 68 \div 100 = 0.68$

  $50\% = 50 \div 100 = \dfrac{50}{100} = \dfrac{1}{2}$

Complete the table. Show your work.

	Fraction (in simplest form)	Decimal	Percent
①	$\dfrac{13}{200}$	$\dfrac{13}{200} = \dfrac{}{1000} =$	$\dfrac{13}{200} \times 100\% =$
②		0.88	0.88 x
③			35%
④	$2\dfrac{4}{25}$		

Put each group of numbers in order from greatest to least.

⑤ 0.53 $\dfrac{9}{25}$ 85%   _____

⑥ $\dfrac{29}{40}$ 0.73   70.5%   _____

⑦ $1\dfrac{1}{2}$ 123%   1.15   _____

ISBN: 978-1-897164-21-1

Solve the problems. Show your work.

⑧ 30% of $15	⑨ 25% of 400 cm
⑩ 145% of 20 kg	⑪ 7.5% of 80 L
⑫ What percent is 15 out of 24?	⑬ What percent is 8 out of 2.5?
⑭ 120 is 80% of what number?	⑮ 48 is 60% of what number?

Fill in the blanks.

⑯ _____ % is 6 out of 150.

⑰ _____ is 25% of 82.4.

⑱ 75.2% of 200 is _____ .

⑲ _____ % of 400 is 64.

⑳ 20% of _____ is 20.

㉑ 90% is _____ out of 80.

㉒ If Uncle Tim sells 45% of a box of oranges,

_____ oranges are sold in all.

㉓ If there are 36 rotten oranges in a box,

_____ % of the oranges are rotten.

㉔ If Mrs. Winter buys a box of oranges and is given a 10% discount,

a.   she can save $ _____ .

b.   she needs to pay $ _____ .

㉕ If Mr. Smith buys a box of oranges at its regular price and needs to pay 15% delivery charge,

a.   the delivery charge is $ _____ .

b.   he needs to pay $ _____ in all.

Equations

1. Write in algebraic expression.

   The product of a number and 5, plus 8.
   Expression: $5n + 8$

2. Evaluate the expression when $m = 6$ and $n = 3$.

   $2mn - \dfrac{m}{n}$ ⟵ mn means $m \times n$; $\dfrac{m}{n}$ means $m \div n$

   $= 2\,(6)\,(3) - \dfrac{(6)}{(3)}$

   $= 36 - 2$

   $= 34$

3. Simplify the expression.

   $6p + 4q - 2p + 10q$ ⟵ 2 like terms for p; 2 like terms for q

   $= 6p - 2p + 4q + 10q$

   $= 4p + 14q$

4. Solve the equation.

   $n + 10 - 5 = 26 - 3$

   $n + 5 = 23$

   $n + 5 - 5 = 23 - 5$

   $n = 18$

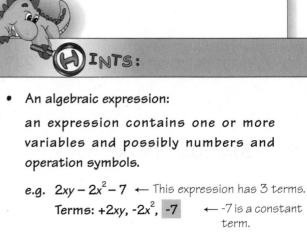

HINTS:

- An algebraic expression:

  an expression contains one or more variables and possibly numbers and operation symbols.

  e.g. $2xy - 2x^2 - 7$ ⟵ This expression has 3 terms.
  Terms: $+2xy$, $-2x^2$, -7 ⟵ -7 is a constant term.

- The Distributive Property:

  $a\,(b + c) = ab + ac$

  e.g. $6x\,(y + z) = 6xy + 6xz$

- Terms that have the same variable with the same exponent are called like terms; otherwise, they are unlike terms.

  e.g. $3x^2$ and $4x^2$ ⟵ like terms
  $3x^2$ and $9x$ ⟵ unlike terms

Write in algebraic expressions.

① The product of a number and 4 _____

② A number divided by 8 _____

③ Half of a number _____

④ 7 less than a number; then divided by 2 _____

⑤ Double a number; then plus 3 _____

⑥ 4 divided by a number decreased by 5 _____

⑦ 0.35 of a number decreased by 2 _____

⑧ A number squared; then increased by 7 _____

Write each expression in words.

⑨ $\dfrac{9}{x}$ _____

⑩ $5n$ _____

⑪ $4x + 1$ _____

⑫ $6(y - 2)$ _____

Evaluate each expression when $a = 3$, $b = -2$, and $c = 1$.

⑬ $6a - 3b$

$= 6(\quad) - 3(\quad)$

$=$ _____

$=$ _____

⑭ $20 - 7ab$

$= 20 - 7(\quad)(\quad)$

$=$ _____

$=$ _____

⑮ $-3b + 2c$

$= -3(\quad) + 2(\quad)$

$=$ _____

$=$ _____

⑯ $8c^3 - 2a^2$

$= 8(\quad)^3 - 2(\quad)^2$

$=$ _____

$=$ _____

Group the like terms in the correct circles.

⑰ $2x^2$ $5xy$ $-2x^2$

   3 $9x^2$ -4

   xy 8 $-2xy$

   0 $3xy$ x^2

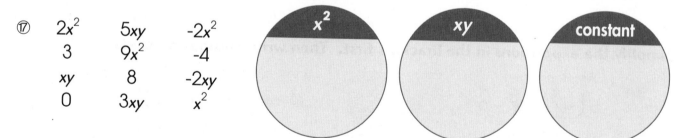

x^2 xy constant

List and write the number of terms for each expression.

⑱ $5 + 2x^2 - 6x + 3xy$ _____ ; ◯ term(S)

⑲ $3x^2 + 6xy - 3x + y^2 + 9$ _____ ; ◯ term(S)

⑳ $h^2 - h$ _____ ; ◯ term(S)

㉑ $-4e - 9e^2 + 7e^3 + 2$ _____ ; ◯ term(S)

㉒ $123abc$ _____ ; ◯ term(S)

ISBN: 978-1-897164-21-1

Use shapes to group the like terms. Then simplify the expressions.

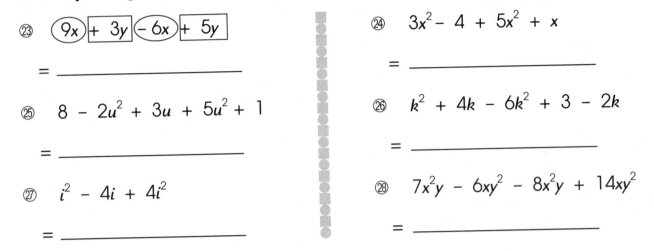

㉓ $\left(9x\right) + \boxed{3y} \left(-6x\right) + \boxed{5y}$

= _____

㉔ $3x^2 - 4 + 5x^2 + x$

= _____

㉕ $8 - 2u^2 + 3u + 5u^2 + 1$

= _____

㉖ $k^2 + 4k - 6k^2 + 3 - 2k$

= _____

㉗ $i^2 - 4i + 4i^2$

= _____

㉘ $7x^2y - 6xy^2 - 8x^2y + 14xy^2$

= _____

Draw arrows to show the application of distributive property. Then write without brackets.

㉙ $4(a+b)$

= _____

㉚ $2(c+d)$

= _____

㉛ $m(m+n)$

= _____

㉜ $3x(x+8)$

= _____

㉝ $5x(y+2)$

= _____

㉞ $3(2x^2-6)$

= _____

㉟ $9b(b-a+4)$

= _____

㊱ $5(6x-7+x^2)$

= _____

㊲ $4(-9+6m+n)$

= _____

Simplify the expressions in the brackets first. Then write without brackets.

㊳ $5(4x^2+3x^2-6)$

= _____

= _____

㊴ $e(6e-7e+1)$

= _____

= _____

㊵ $3(2y+9y^2-4y)$

= _____

= _____

㊶ $4h(h^2-2h-2h^2)$

= _____

= _____

ISBN: 978-1-897164-21-1

Solve the equations.

㊷ $d - 4 = 11$ $d - 4 + 4 = 11 + 4$ $d =$	㊸ $n + 4 = 24$	㊹ $s - 16 = 2$
㊺ $4i = 12$	㊻ $\frac{k}{9} = 10$	㊼ $14n = 28$
㊽ $\frac{3}{4}y = 24$	㊾ $\frac{4}{3}e = 32$	㊿ $4x + 4x = 5.3 - 0.5$
(51) $5n - 3n = 20 + 4$	(52) $6a + 5a = (33)(3)$	(53) $\frac{i}{3} + \frac{2i}{3} = 4$

Set up the equation for each word problem; then solve the unknown.

(54) A rectangle has an area of 165 cm^2. Find its width if the length is 11 cm.

The width of the rectangle is ________ cm.

(55) John has 2 boxes of marbles. Each box contains the same number of marbles. If he gets 5 more marbles, he will have 39 marbles in all. How many marbles are there in each box?

(56) Mrs. Winter pays $45 for the tax of a ring. If the tax rate is 15%, how much does the ring cost?

ISBN: 978-1-897164-21-1

Review

Circle the letter which represents the correct answer in each problem.

① 108 is the value of

    A. 2^2 B. 3^3 C. $2^2 \times 3^3$ D. $2^3 \times 3^2$

② The value of $8 \times 10^3 + 4 \times 10^1 + 3 \times 10^0$ is

    A. 8403   B. 8043   C. 843   D. 8430

③ $\sqrt{174}$ lies between

    A. 11 and 12   B. 12 and 13   C. 13 and 14   D. 14 and 15

④ The greatest common factor of 18, 24, and 36 is

    A. 72   B. 6   C. 12   D. 36

⑤ The least common multiple of 12, 16, and 24 is

    A. 48   B. 4   C. 72   D. 24

⑥ Which of the following is the smallest?

    A. 58%   B. 5.8   C. $\frac{5}{8}$ D. $\frac{5}{8}$ %

⑦ $6\frac{1}{3} + 1\frac{1}{8} \times 1\frac{1}{3}$ is

    A. $9\frac{17}{18}$ B. $5\frac{6}{7}$ C. $6\frac{5}{7}$ D. $7\frac{5}{6}$

⑧ The unit rate for 1 km in 25 min is

    A. 0.04 m/min   B. 40 m/min

    C. 25 m/min   D. 40 km/min

⑨ Which of the following is a like term of $16xy^2$?

    A. $8x^2y$ B. $16x^2y$ C. $8xy^2$ D. $16xy$

Find the answers.

⑩ (+5) + (-3) = _____

⑪ (+6) – (-8) = _____

⑫ (-3) (-6) = _____

⑬ (-9) ÷ (+3) = _____

⑭ (+6) – (-7) x (-5) = _____

⑮ (+2) + (-6) ÷ (-3) = _____

⑯ 4.26 ÷ (2.4 ÷ 2) = _____

⑰ 0.55 x 20 ÷ 4   = _____

⑱ 13.31 – 9.88 ÷ 4 = _____

⑲ 4 x 0.12 + 3 x 0.8 = _____

Find the answers. Show the steps. Reduce the answers to lowest terms.

⑳ $1\frac{1}{3} \times (\frac{1}{4} + \frac{7}{8})$

㉑ $7\frac{2}{7} - 1\frac{1}{2} \times 4\frac{2}{7}$

㉒ $1\frac{7}{18} - \frac{5}{6} \div 15$

Look at the shapes. Fill in the blanks.

㉓
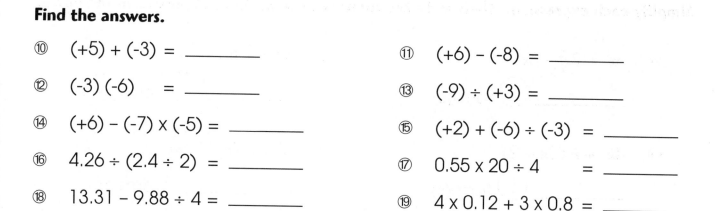

a. _____ out of 20 shapes are triangles; _____ % of the shapes are triangles.

b. _____ out of 20 shapes are hearts; _____ % of the shapes are hearts.

c. The ratio of the number of triangles to the number of shapes is _____ .

Answer the questions.

㉔ What is the value of 30% of 224? _____

㉕ What is the value of 120% of 50? _____

㉖ What percent is 75 out of 500? _____

㉗ 32% of a number is 25.6. What is the number? _____

㉘ 40% of a number is 50. What is the number? _____

ISBN: 978-1-897164-21-1

Review

Simplify each expression. Then write the number of terms in each answer in the circle.

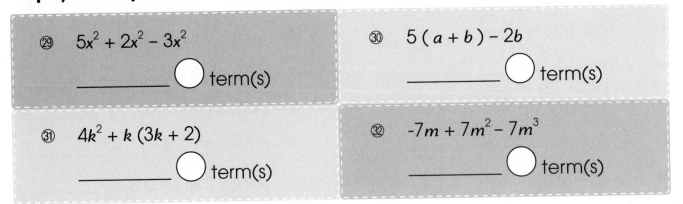

㉙ $5x^2 + 2x^2 - 3x^2$

_____ ◯ term(s)

㉚ $5(a+b) - 2b$

_____ ◯ term(s)

㉛ $4k^2 + k(3k+2)$

_____ ◯ term(s)

㉜ $-7m + 7m^2 - 7m^3$

_____ ◯ term(s)

Solve the equations. Show all the steps.

㉝ $5 + n = 17$	㉞ $16x = 80$	㉟ $\dfrac{k}{4} = 12$
㊱ $7m = 15 + 41$	㊲ $9y + 12y = 63$	㊳ $19e - 7e = 90 + 66$

Solve the problems.

㊴ 75% of the customers in the store yesterday were women. If there were 584 customers yesterday, how many female customers were there?

㊵ Susan has a bag of candies and shares it equally with Tim. If Susan eats 3 candies, she will have 6 candies left. How many candies are there in the bag? (Use an equation to solve the problem.)

㊶ The parallelogram has an area of 225 cm². Find the height of the parallelogram if its length is 5 cm.

ISBN: 978-1-897164-21-1

Overview

Section I enables students to strengthen their basic mathematical skills. In this section, these skills are built upon and applied in various day-to-day applications such as money, calculations of discount and tax, and temperature measurements.

Geometry units cover perimeters, areas, volumes, angles, and the construction of shapes. Moreover, students will be introduced to coordinates and transformations.

Data Management topics include statistics and various graphic representations of data. Students will continue to examine probability through spinners and coins.

Algebraic expressions and equations are also studied in this section. Students will learn how to write and solve equations. Furthermore, they will learn how to graph simple equations.

1 Number Theory

Power	- a product of equal factors
Base (in a power)	- the factor repeated in a power
Exponent (in a power)	- the number of times the base occurs as a factor

base   exponent

e.g. $2 \times 2 \times 2 \times 2 \times 2 = 2^5 = 32$; 2^5 is called a power.

In writing, it is two to the fifth power.

Square number	- the product of a number multiplied by itself
Square root	- one of the two equal factors of a number

Follow Tony's method to write each product as a power and state the base an exponent of each power.

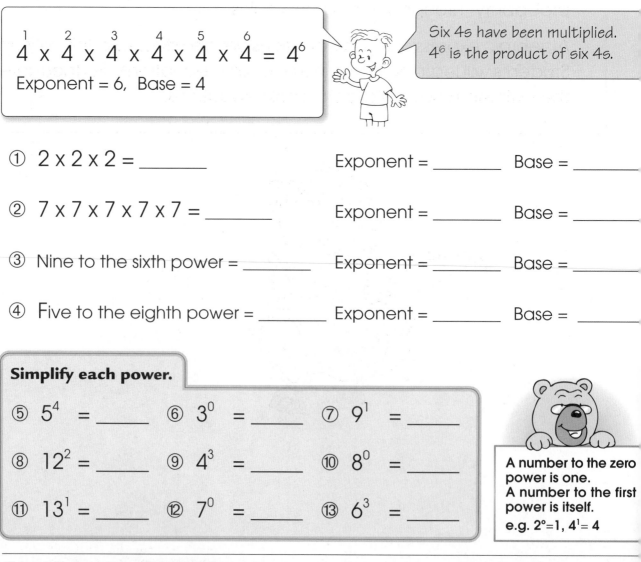

$$\overset{1}{4} \times \overset{2}{4} \times \overset{3}{4} \times \overset{4}{4} \times \overset{5}{4} \times \overset{6}{4} = 4^6$$

Exponent = 6, Base = 4

Six 4s have been multiplied. 4^6 is the product of six 4s.

① $2 \times 2 \times 2 =$ _____   Exponent = _____ Base = _____

② $7 \times 7 \times 7 \times 7 \times 7 =$ _____   Exponent = _____ Base = _____

③ Nine to the sixth power = _____   Exponent = _____ Base = _____

④ Five to the eighth power = _____ Exponent = _____ Base = _____

Simplify each power.

⑤ 5^4 = _____   ⑥ 3^0 = _____   ⑦ 9^1 = _____

⑧ 12^2 = _____   ⑨ 4^3 = _____   ⑩ 8^0 = _____

⑪ 13^1 = _____   ⑫ 7^0 = _____   ⑬ 6^3 = _____

A number to the zero power is one.
A number to the first power is itself.
e.g. $2^0 = 1$, $4^1 = 4$

ead what Tony says. Then write each product as a power of 10.

There are about 10^4 people in the parade.

4 tens are multiplied.

10^4 = 10 x 10 x 10 x 10

= 10 000 → 1 is followed by 4 zeros.

⑭ 10 x 10 = _____ ⑮ 10 x 10 x 10 x 10 x 10 = _____

⑯ 10 x 10 x 10 = _____ ⑰ 10 x 10 x 10 x 10 = _____

⑱ one million = _____ ⑲ one hundred thousand = _____

⑳ 10 x 10 x 10 x 10 x 10 x 10 x 10 = _____

ollow Tony's method to express each number as powers of 10.

9 806 = 9000 + 800 + 0 + 6

= 9 x 1000 + 8 x 100 + 0 x 10 + 6 x 1

= $9 \times 10^3 + 8 \times 10^2 + 0 \times 10^1 + 6 \times 10^0$

There are 9 806 people in the parade.

㉑ 1 463 = _____ x 10^3 + _____ x 10^2 + _____ x 10^1 + _____ x 10^0

㉒ 3 075 = _____

㉓ 159 = _____

㉔ 42 062 = _____

Find the value.

㉕ $2 \times 10^4 + 1 \times 10^3 + 4 \times 10^2 + 3 \times 10^1 + 2 \times 10^0$ = _____

㉖ $3 \times 10^4 + 5 \times 10^2 + 9 \times 10^0$ = _____

㉗ $6 \times 10^5 + 3 \times 10^2 + 3 \times 10^1 + 2 \times 10^0$ = _____

㉘ $4 \times 10^4 + 3 \times 10^2 + 5 \times 10^0$ = _____

Complete the factor trees and write each number as a product of prime factors.

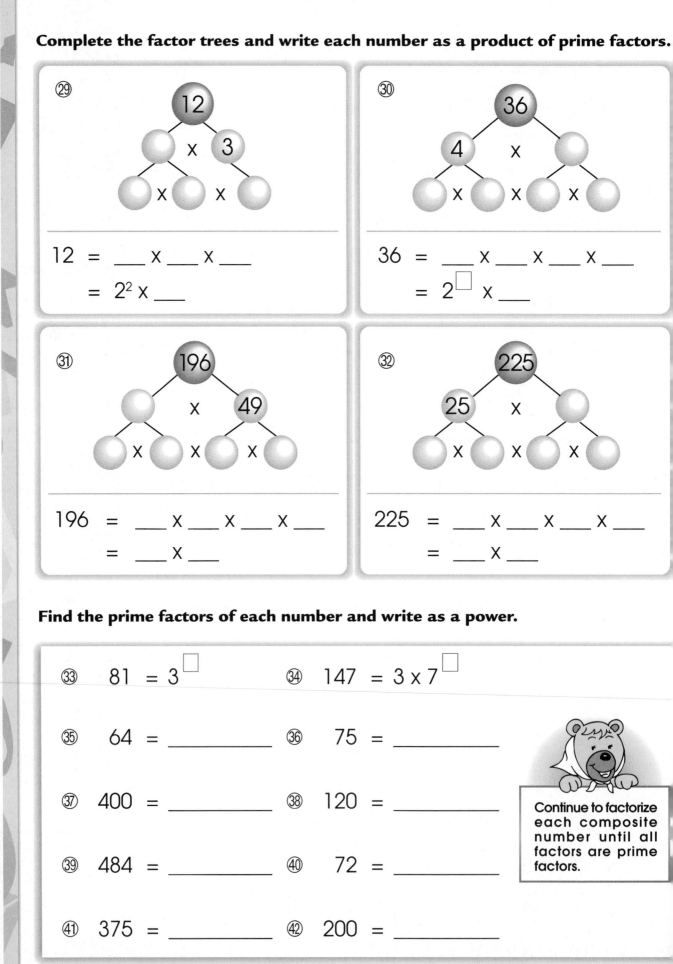

㉙

12 = ___ x ___ x ___

   = 2^2 x ___

㉚

36 = ___ x ___ x ___ x ___

   = $2^\square$ x ___

㉛

196 = ___ x ___ x ___ x ___

    = ___ x ___

㉜

225 = ___ x ___ x ___ x ___

    = ___ x ___

Find the prime factors of each number and write as a power.

㉝ 81 = $3^\square$

㉞ 147 = $3 \times 7^\square$

㉟ 64 = _____

㊱ 75 = _____

㊲ 400 = _____

㊳ 120 = _____

㊴ 484 = _____

㊵ 72 = _____

Continue to factorize each composite number until all factors are prime factors.

㊶ 375 = _____

㊷ 200 = _____

ISBN: 978-1-897164-21-1

Follow Tony's method to find the square root of each number.

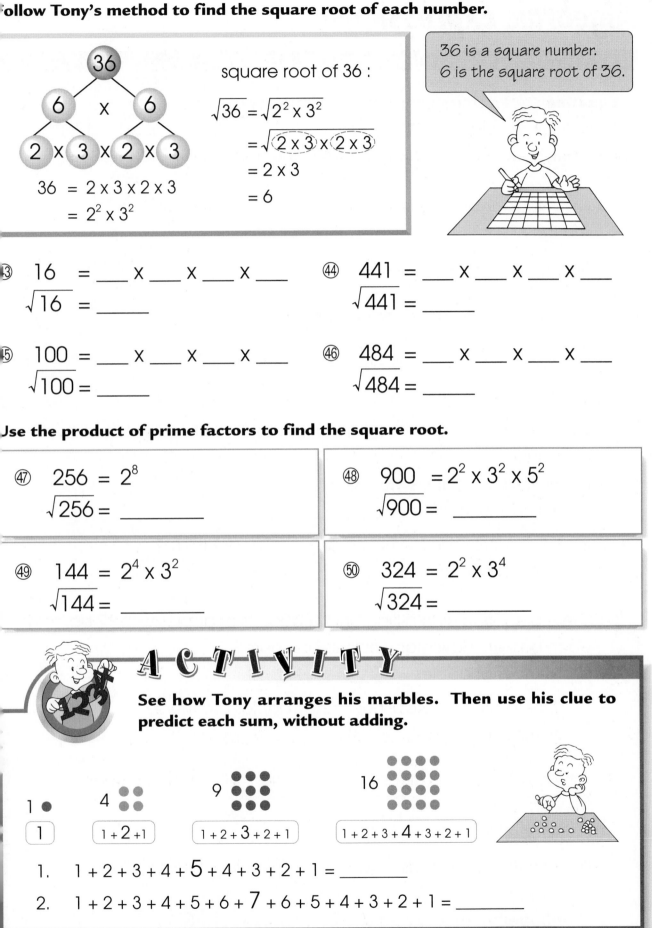

square root of 36 :

$$\sqrt{36} = \sqrt{2^2 \times 3^2}$$
$$= \sqrt{(2 \times 3) \times (2 \times 3)}$$
$$= 2 \times 3$$
$$= 6$$

36 is a square number.
6 is the square root of 36.

$36 = 2 \times 3 \times 2 \times 3$
$\quad = 2^2 \times 3^2$

43 $16 = \underline{} \times \underline{} \times \underline{} \times \underline{}$

$\sqrt{16} = \underline{}$

44 $441 = \underline{} \times \underline{} \times \underline{} \times \underline{}$

$\sqrt{441} = \underline{}$

45 $100 = \underline{} \times \underline{} \times \underline{} \times \underline{}$

$\sqrt{100} = \underline{}$

46 $484 = \underline{} \times \underline{} \times \underline{} \times \underline{}$

$\sqrt{484} = \underline{}$

Use the product of prime factors to find the square root.

47 $256 = 2^8$

$\sqrt{256} = \underline{}$

48 $900 = 2^2 \times 3^2 \times 5^2$

$\sqrt{900} = \underline{}$

49 $144 = 2^4 \times 3^2$

$\sqrt{144} = \underline{}$

50 $324 = 2^2 \times 3^4$

$\sqrt{324} = \underline{}$

ACTIVITY

See how Tony arranges his marbles. Then use his clue to predict each sum, without adding.

1 •
$\boxed{1}$

4
$\boxed{1 + 2 + 1}$

9
$\boxed{1 + 2 + 3 + 2 + 1}$

16
$\boxed{1 + 2 + 3 + 4 + 3 + 2 + 1}$

1. $1 + 2 + 3 + 4 + 5 + 4 + 3 + 2 + 1 = \underline{}$

2. $1 + 2 + 3 + 4 + 5 + 6 + 7 + 6 + 5 + 4 + 3 + 2 + 1 = \underline{}$

2 Algebraic Expressions

Equation - a mathematical sentence with an equal sign (=)

Follow Tony's method to write the expressions and answer the questions.

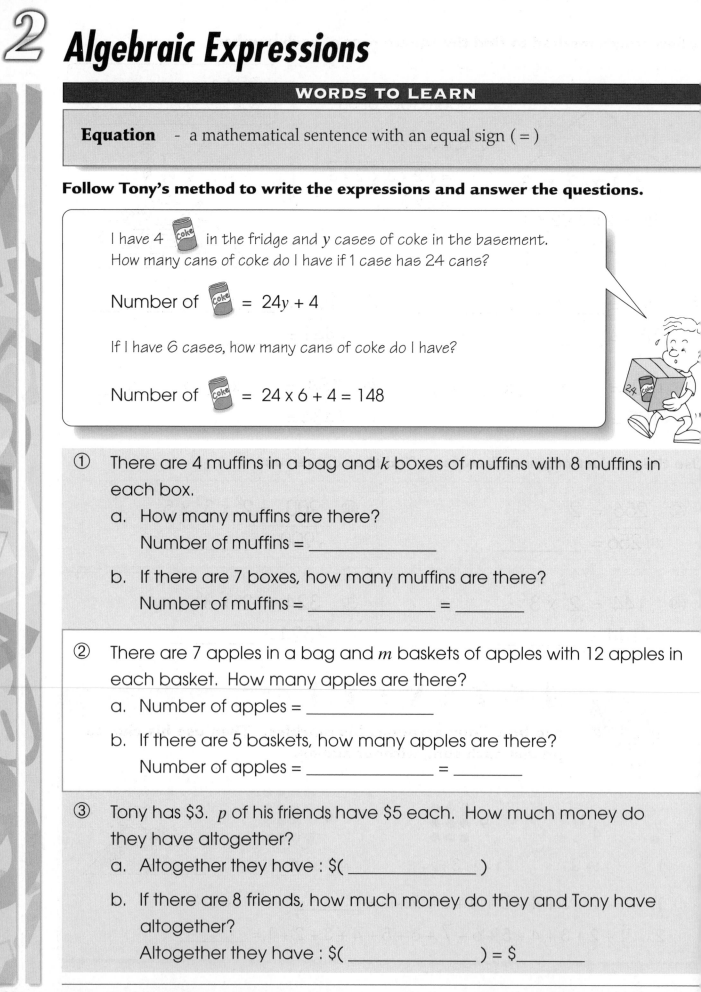

I have 4 [coke] in the fridge and y cases of coke in the basement. How many cans of coke do I have if 1 case has 24 cans?

Number of [coke] = $24y + 4$

If I have 6 cases, how many cans of coke do I have?

Number of [coke] = $24 \times 6 + 4 = 148$

① There are 4 muffins in a bag and k boxes of muffins with 8 muffins in each box.
   a. How many muffins are there?
      Number of muffins = _____

   b. If there are 7 boxes, how many muffins are there?
      Number of muffins = _____ = _____

② There are 7 apples in a bag and m baskets of apples with 12 apples in each basket. How many apples are there?
   a. Number of apples = _____

   b. If there are 5 baskets, how many apples are there?
      Number of apples = _____ = _____

③ Tony has $3. p of his friends have $5 each. How much money do they have altogether?
   a. Altogether they have : $(_____)

   b. If there are 8 friends, how much money do they and Tony have altogether?
      Altogether they have : $(_____) = $_____

Evaluate each expression when $n = 12$.

④ $3n - 9$ =

⑤ $6 + 7n$ =

⑥ $n \div 3 + 12$ =

⑦ $81 - n \div 2$ =

Evaluate each expression when $x = 3$ and $y = 16$.

⑧ $2x + y =$ _____

⑨ $3y - 4x =$ _____

⑩ $y \div 4 + 2x =$ _____

⑪ $3x - y \div 8 =$ _____

⑫ $y - 2x + 6 =$ _____

⑬ $y \div 2 + x \div 3 =$ _____

> Substitute the numbers for the unknowns, but remember the order of operations.

Follow Tony's method to solve the equations.

Addition equation	Subtraction equation
$x + 8 = 15$	$y - 6 = 10$
$x + 8 - 8 = 15 - 8$	$y - 6 + 6 = 10 + 6$
$x = 7$	$y = 16$
Subtract 8 from both sides.	Add 6 to both sides.

> To solve an addition equation, subtract the same number from both sides of the equation.

> To solve a subtraction equation, add the same number to both sides of the equation.

⑭ $k + 5 = 19$ $k =$ _____

⑮ $p - 0.7 = 8$ $p =$ _____

⑯ $v + 2.4 = 6$ $v =$ _____

⑰ $u + 1.3 = 2.1$ $u =$ _____

⑱ $9.8 + a = 10.2$ $a =$ _____

⑲ $p - 4 = 2.3$ $p =$ _____

⑳ $1.1 + m = 1.1$ $m =$ _____

㉑ $n - 0.8 = 0.8$ $n =$ _____

㉒ $z - 32 = 4$ $z =$ _____

㉓ $b - 3 = \dfrac{1}{2}$ $b =$ _____

㉔ $q - \dfrac{1}{2} = \dfrac{1}{4}$ $q =$ _____

㉕ $r + \dfrac{1}{5} = \dfrac{3}{5}$ $r =$ _____

Follow Helen's method to solve the equations.

Multiplication equation	Division equation
$3y = 18$	$x \div 6 = 2$
$3y \div 3 = 18 \div 3$	$x \div 6 \times 6 = 2 \times 6$
$y = 6$	$x = 12$
Divide both sides by 3.	Multiply both sides by 6.

To solve a multi-plication equation, divide both sides of the equation by the same number.

To solve a division equation, multiply both sides of the equation by the same number.

㉖ $4n = 40$ $\qquad n = \underline{\hspace{1.5cm}}$

㉗ $0.2a = 3$ $\qquad a = \underline{\hspace{1.5cm}}$

㉘ $\dfrac{1}{2}q = 11$ $\qquad q = \underline{\hspace{1.5cm}}$ ㉙ $p \times 0.4 = 2$ $\qquad p = \underline{\hspace{1cm}}$

㉚ $\dfrac{v}{5} = 0.2$ $\qquad v = \underline{\hspace{1.5cm}}$ ㉛ $z \div 7 = 0.5$ $\qquad z = \underline{\hspace{1cm}}$

㉜ $0.9\,s = 0$ $\qquad s = \underline{\hspace{1.5cm}}$ ㉝ $0.8u = 0.2$ $\qquad u = \underline{\hspace{1cm}}$

㉞ $\dfrac{b}{5} = 5$ $\qquad b = \underline{\hspace{1.5cm}}$ ㉟ $t \div 1.2 = 1$ $\qquad t = \underline{\hspace{1cm}}$

Find the ordered pairs and draw the graphs. Then use the graphs to find t **values of y.**

When $x = 1$, then $y = 1 + 1 = 2$.

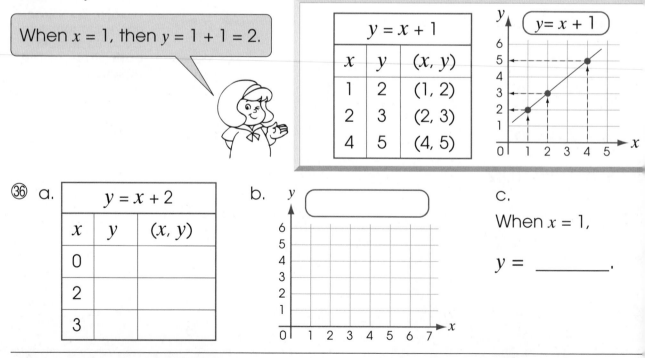

$y = x + 1$		
x	y	(x, y)
1	2	(1, 2)
2	3	(2, 3)
4	5	(4, 5)

㊱ a.

$y = x + 2$		
x	y	(x, y)
0		
2		
3		

b.

c.

When $x = 1$,

$y = \underline{\hspace{2cm}}$.

ISBN: 978-1-897164-21-1

㊲ a.

y = x − 1		
x	y	(x, y)
1		
3		
5		

b.

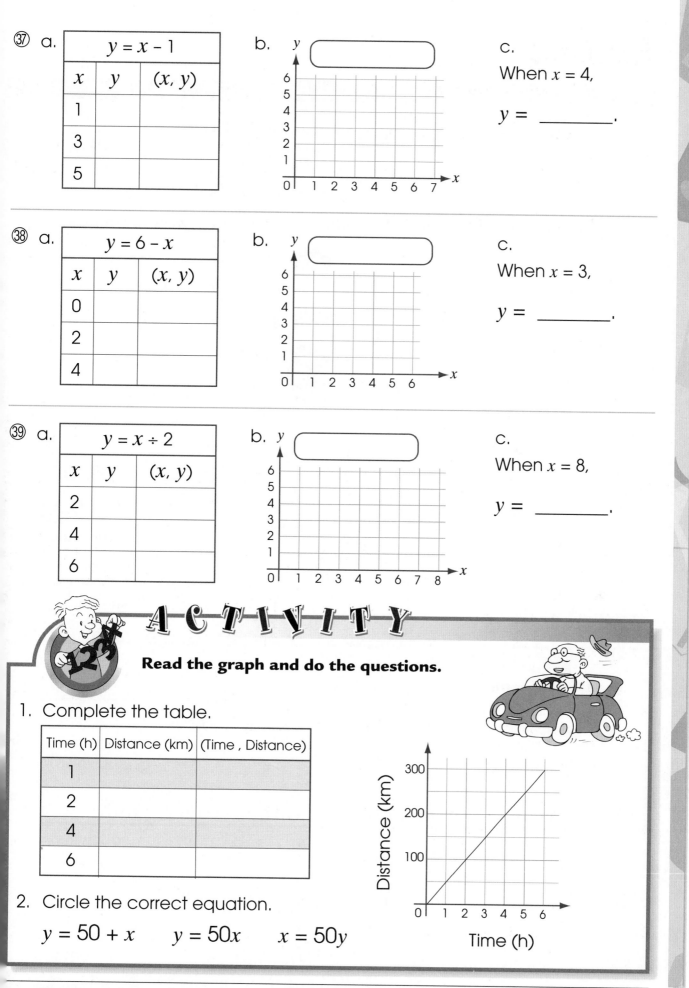

c.

When x = 4,

y = _____.

㊳ a.

y = 6 − x		
x	y	(x, y)
0		
2		
4		

b.

c.

When x = 3,

y = _____.

㊴ a.

y = x ÷ 2		
x	y	(x, y)
2		
4		
6		

b.

c.

When x = 8,

y = _____.

ACTIVITY

Read the graph and do the questions.

1. Complete the table.

Time (h)	Distance (km)	(Time , Distance)
1		
2		
4		
6		

2. Circle the correct equation.

$y = 50 + x$ $y = 50x$ $x = 50y$

ISBN: 978-1-897164-21-1

3 Fractions

Reciprocal - either of a pair of numbers whose product is 1

e.g. $\dfrac{4}{3}$ is the reciprocal of $\dfrac{3}{4}$.

Follow Dave's method to find each product. Write the answers in simplest form.

$$3\frac{1}{3} \times 2\frac{1}{4} = \frac{\overset{5}{10}}{\underset{1}{3}} \times \frac{\overset{3}{9}}{\underset{2}{4}}$$

$$= \frac{5 \times 3}{2}$$

$$= \frac{15}{2} = 7\frac{1}{2}$$

1st Change the mixed numbers to improper fractions.

2nd Simplify by the common factors.

3rd Multiply.

4th Write in simplest form.

① $2\dfrac{1}{3} \times 1\dfrac{1}{2} =$

② $5\dfrac{1}{2} \times \dfrac{4}{11} =$

③ $2\dfrac{5}{6} \times 7\dfrac{1}{2} =$

④ $4 \times 3\dfrac{5}{8} =$

⑤ $3\dfrac{3}{7} \times 14 =$

⑥ $7\dfrac{1}{3} \times 4\dfrac{1}{2} =$

You can write the whole number as an improper fraction with denominator of 1, e.g. $4 = \dfrac{4}{1}$

Solve the problems.

⑦ $\dfrac{3}{4}$ of 40 students went climbing. How many students went climbing?

_____ students

⑧ Tony climbed $3\dfrac{1}{2}$ m/min. How far could he climb in $2\dfrac{1}{4}$ min?

_____ m

⑨ Tony brought 2L of water with him. He drank $\dfrac{7}{8}$ of it. How much did he drink?

_____ L

ISBN: 978-1-897164-21-1

ollow Elaine's method to do the division. Write the answers in simplest form.

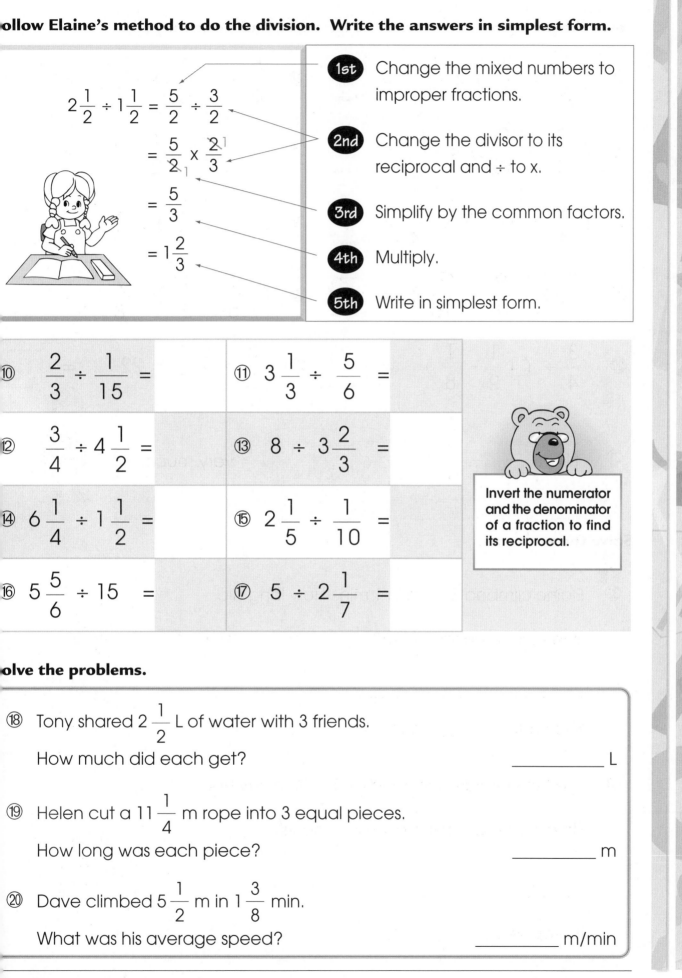

$$2\frac{1}{2} \div 1\frac{1}{2} = \frac{5}{2} \div \frac{3}{2}$$

$$= \frac{5}{2} \times \frac{2}{3}$$

$$= \frac{5}{3}$$

$$= 1\frac{2}{3}$$

1st Change the mixed numbers to improper fractions.

2nd Change the divisor to its reciprocal and ÷ to x.

3rd Simplify by the common factors.

4th Multiply.

5th Write in simplest form.

⑩ $\dfrac{2}{3} \div \dfrac{1}{15} =$

⑪ $3\dfrac{1}{3} \div \dfrac{5}{6} =$

⑫ $\dfrac{3}{4} \div 4\dfrac{1}{2} =$

⑬ $8 \div 3\dfrac{2}{3} =$

Invert the numerator and the denominator of a fraction to find its reciprocal.

⑭ $6\dfrac{1}{4} \div 1\dfrac{1}{2} =$

⑮ $2\dfrac{1}{5} \div \dfrac{1}{10} =$

⑯ $5\dfrac{5}{6} \div 15 =$

⑰ $5 \div 2\dfrac{1}{7} =$

olve the problems.

⑱ Tony shared $2\dfrac{1}{2}$ L of water with 3 friends.

How much did each get? _____ L

⑲ Helen cut a $11\dfrac{1}{4}$ m rope into 3 equal pieces.

How long was each piece? _____ m

⑳ Dave climbed $5\dfrac{1}{2}$ m in $1\dfrac{3}{8}$ min.

What was his average speed? _____ m/min

ISBN: 978-1-897164-21-1

Solve the problems. Then write the letters to find what Tony says.

㉑ $\left(\dfrac{1}{5} + \dfrac{2}{5}\right) \times \dfrac{5}{7} = \boxed{i}$ ㉒ $\dfrac{5}{6} \div \left(\dfrac{1}{12} + \dfrac{5}{12}\right) = \boxed{e}$

㉓ $\left(\dfrac{3}{10} + \dfrac{9}{10}\right) \div \dfrac{4}{5} = \boxed{g}$ ㉔ $8 \div \dfrac{4}{7} \times \dfrac{1}{2} = \boxed{m}$

㉕ $\left(\dfrac{1}{4} + 1\dfrac{1}{2}\right) \times \dfrac{1}{14} = \boxed{c}$ ㉖ $\left(4 - 2\dfrac{1}{3}\right) \times \dfrac{3}{7} = \boxed{a}$

㉗ $\dfrac{2}{3} \times \left(\dfrac{9}{16} - \dfrac{1}{16}\right) = \boxed{n}$ ㉘ $\left(1\dfrac{1}{5} - \dfrac{1}{2}\right) \div \dfrac{7}{8} = \boxed{i}$

㉙ $\dfrac{3}{4} \div \left(1\dfrac{1}{2} - \dfrac{1}{8}\right) = \boxed{l}$ ㉚ $\dfrac{3}{5} \div 2\dfrac{1}{5} \times 22 = \boxed{b}$

㉛ I like $\boxed{\dfrac{1}{8}}$ $\boxed{\dfrac{6}{11}}$ $\boxed{\dfrac{3}{7}}$ $\boxed{7}$ $\boxed{6}$ $\boxed{\dfrac{4}{5}}$ $\boxed{\dfrac{1}{3}}$ $\boxed{1\dfrac{1}{2}}$ very much.

Solve the problems.

㉜ Elaine climbed $3\dfrac{1}{3}$ m in 1 min. How long did she take to climb $7\dfrac{1}{2}$ m?

_____

She took _____ min.

㉝ The temperature dropped by $1\dfrac{1}{5}$ °C every hour. How many °C did it drop in $3\dfrac{1}{3}$ hours?

_____

It dropped _____ °C.

ISBN: 978-1-897164-21-1

㉞ Helen spilt $\frac{2}{3}$ of a $2\frac{1}{2}$ L bottle of water. How much water did she spill?

____

She spilt ________ L of water.

㉟ Dave shared $\frac{3}{4}$ kg of jelly beans with 3 friends. How many grams of beans did each get?

____

Each got ________ grams of jelly beans.

㊱ A bottle of water was $\frac{7}{8}$ full. Tony drank $\frac{2}{5}$ of it. How much water was left?

____

________ of a bottle of water was left.

ACTIVITY

Write the answers as fractions in simplest form.

1.

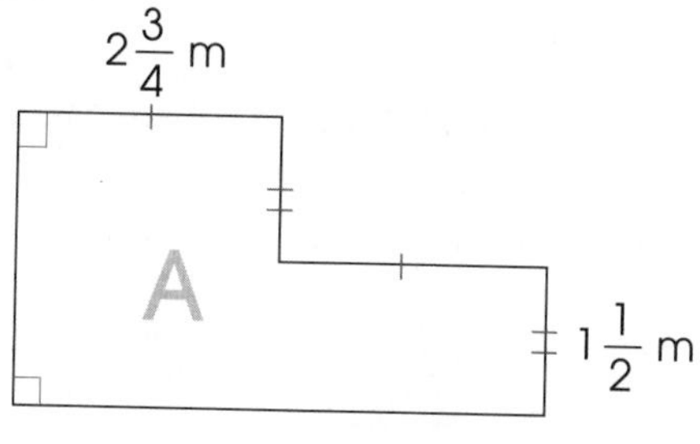

$2\frac{3}{4}$ m
A
$1\frac{1}{2}$ m

Area = ________ m^2

2.

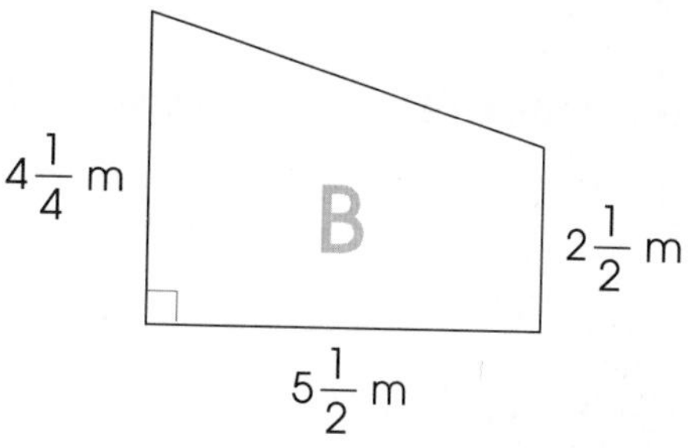

$4\frac{1}{4}$ m
B
$2\frac{1}{2}$ m
$5\frac{1}{2}$ m

Area = ________ m^2

3. How many times is Shape B bigger than Shape A? ☐ times

ISBN: 978-1-897164-21-1

4 Percents

Discount - the difference between the regular price and the sale price of an item

Sales tax - money paid to the government on things we buy

Express each fraction as a percent. Round to the nearest hundredth, if necessary.

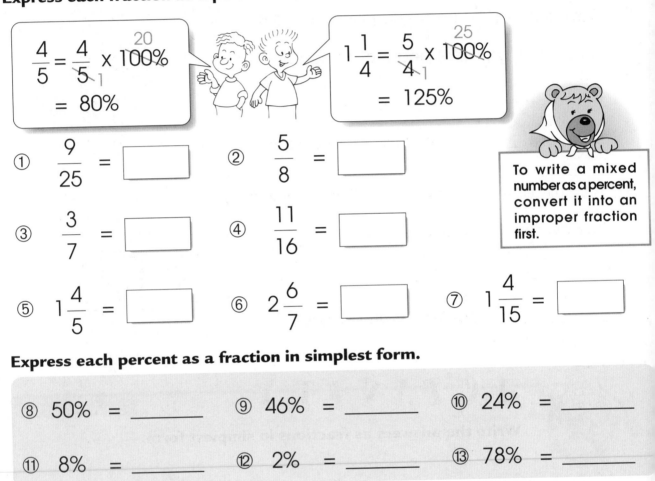

$$\frac{4}{5} = \frac{4}{5} \times \overset{20}{\cancel{100}}\%_1$$
$$= 80\%$$

$$1\frac{1}{4} = \frac{5}{4} \times \overset{25}{\cancel{100}}\%_1$$
$$= 125\%$$

To write a mixed number as a percent, convert it into an improper fraction first.

① $\frac{9}{25}$ = ☐

② $\frac{5}{8}$ = ☐

③ $\frac{3}{7}$ = ☐

④ $\frac{11}{16}$ = ☐

⑤ $1\frac{4}{5}$ = ☐

⑥ $2\frac{6}{7}$ = ☐

⑦ $1\frac{4}{15}$ = ☐

Express each percent as a fraction in simplest form.

⑧ 50% = _____

⑨ 46% = _____

⑩ 24% = _____

⑪ 8% = _____

⑫ 2% = _____

⑬ 78% = _____

Find the percent of the shaded parts in each shape. Round to the nearest hundredth if necessary.

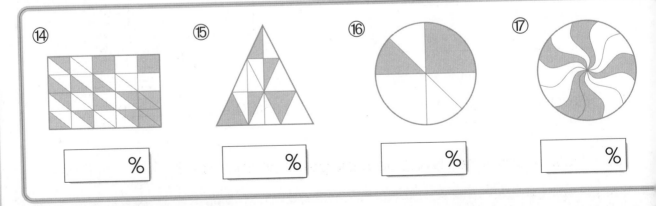

⑭ ☐ %

⑮ ☐ %

⑯ ☐ %

⑰ ☐ %

ISBN: 978-1-897164-21-1

Follow Tony's method to solve the problems.

30.5% of 400 T-shirts are on sale. How many T-shirts are on sale?

$$400 \times 30.5\% = \overset{4}{400} \times \frac{30.5}{100_{1}} = 122$$

122 T-shirts are on sale.

⑱ 15% of 250 = _____

⑲ 16% of 210 = _____

⑳ 32.5% of 500 = _____

㉑ 18.7% of 400 = _____

㉒ 64% of 37.5 = _____

㉓ 25% of 64.8 = _____

㉔ There are 120 skirts in the store. 15% of them are blue, 20% red and the rest black.

    a. How many blue skirts are there?   _____ blue skirts

    b. How many red skirts are there?   _____ red skirts

    c. How many black skirts are there?   _____ black skirts

Read what Dave says. Then help him solve the problems.

90 out of 120 customers in the store are women. What percent of the customers are women?

$$\frac{90}{120} \times 100\% = 75\%$$

㉕ 35 out of 350 _____%

㉖ 75 out of 250 _____%

㉗ $7\frac{1}{2}$ out of 200 _____%

㉘ 6 out of 120 _____%

㉙ 65 out of 250 people in the store have bought something. How many percent of the people have bought things in the store?

_____% of the people have bought things in the store.

See how Tony finds the sale price of the cap. Then help him find the sale prices of the other items. Round to the nearest hundredth, if necessary.

Reg. $15
20% Off
Sale $ 12

Since 100% − 20% = 80%, the sale price is 80% of the regular price.

80% of $15 = 0.8 x $15
= $12

The sale price of the cap is $12.

③⓪ Reg. $41.50
15% Off
Sale $ _____

③① Reg. $82.75
25% Off
Sale $ _____

③② Reg. $23.90
40% Off
Sale $ _____

③⑤ Reg. $34.50
30% Off
Sale $ _____

③③ Reg. $69.99
10% Off
Sale $ _____

③④ Reg. $32.90
35% Off
Sale $ _____

The tax rate is 15%. Help Tony find the total cost of each item. Round to the nearest hundredth, if necessary.

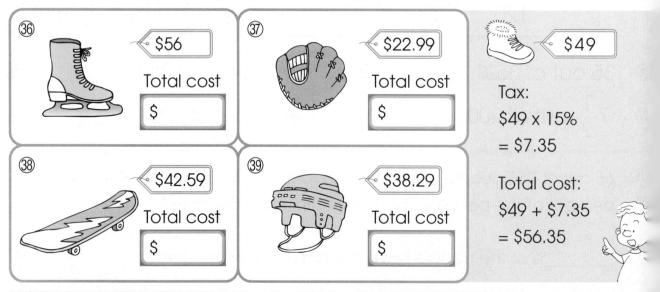

③⑥ $56
Total cost
$ _____

③⑦ $22.99
Total cost
$ _____

$49
Tax:
$49 x 15%
= $7.35

Total cost:
$49 + $7.35
= $56.35

③⑧ $42.59
Total cost
$ _____

③⑨ $38.29
Total cost
$ _____

ISBN: 978-1-897164-21-1

Read what the children say. Then complete the tables.

Regular Price	Discount Rate	Amount of Discount	Sale Price
$20	40%	$8	$12
$90			$58.50
		$11.25	$63.75
$35.85		$7.17	

⑩ $90 row, ⑪ third row, ⑫ $35.85 row

Selling Price	Tax Rate	Amount of Tax	Total Cost
$16	15%	$2.40	$18.40
$40		$2.80	
		$3.54	$33.04
$82.50			$87.45

I have saved $8 out of $20.

Discount rate

$= \dfrac{8}{20} \times 100\%$

$= 40\%$

Tax rate

$= \dfrac{2.4}{16} \times 100\%$

$= 15\%$

Solve the problems. Round the answers to the nearest hundredth, if necessary.

⑯ All items in a shop are 25% off. What is the sale price of a $36 helmet if there is an additional 15% off the reduced price?

$_____

⑰ Helen bought a $46.59 skirt at 30% off. How much did she pay with a tax rate of 9%?

$_____

⑱ Tony bought a shirt for $22.50 and bought a second one at a discount of 50%. How much did he pay for the two shirts with a tax rate of 7%?

$_____

ACTIVITY

The dimensions of Uncle Fred's shop are 18m by 10m, but it will be 10% longer and 10% wider after the renovations. How much bigger will Uncle Fred's shop be after the renovations?

Closed for renovation

His shop will be _____ m² bigger.

ISBN: 978-1-897164-21-1

5 Measurement

WORDS TO LEARN

Perimeter - the distance around the outside of a shape

Area - the number of square units of a surface

Volume - the number of cubic units occupied by an object

Help Tony find the perimeter and area of each shape.

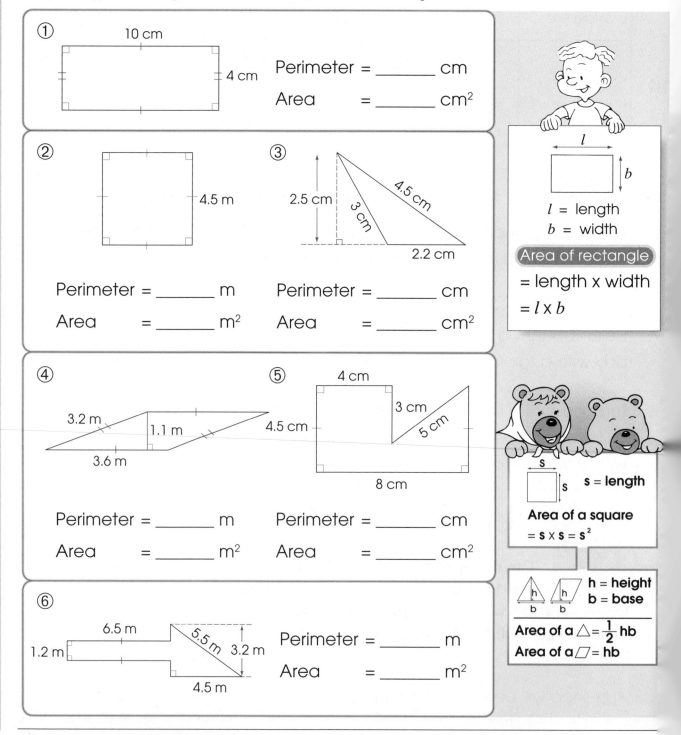

① 10 cm / 4 cm

Perimeter = _____ cm

Area = _____ cm²

② 4.5 m

Perimeter = _____ m

Area = _____ m²

③ 2.5 cm, 3 cm, 4.5 cm, 2.2 cm

Perimeter = _____ cm

Area = _____ cm²

④ 3.2 m, 1.1 m, 3.6 m

Perimeter = _____ m

Area = _____ m²

⑤ 4 cm, 3 cm, 5 cm, 4.5 cm, 8 cm

Perimeter = _____ cm

Area = _____ cm²

⑥ 6.5 m, 1.2 m, 5.5 m, 3.2 m, 4.5 m

Perimeter = _____ m

Area = _____ m²

l = length
b = width

Area of rectangle
= length x width
= l x b

s = length

Area of a square
= s x s = s²

h = height
b = base

Area of a △ = ½ hb
Area of a ▱ = hb

ISBN: 978-1-897164-21-1

Follow Tony's method to find the area of each trapezoid.

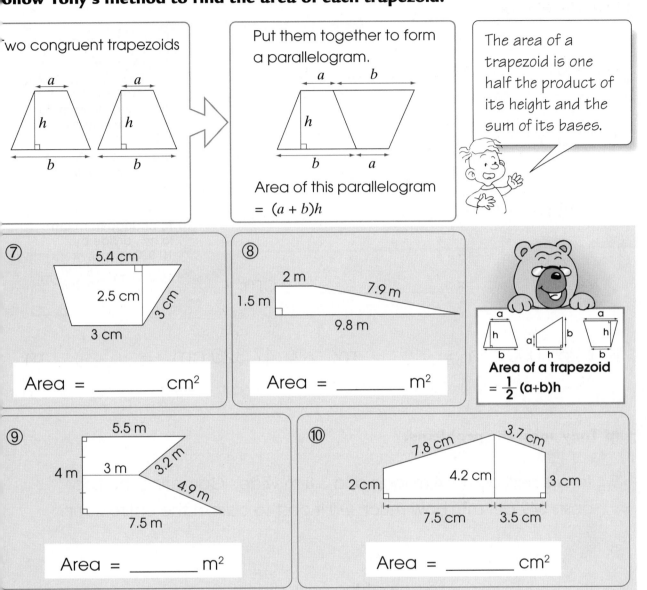

Two congruent trapezoids

Put them together to form a parallelogram.

Area of this parallelogram
= $(a + b)h$

The area of a trapezoid is one half the product of its height and the sum of its bases.

⑦
5.4 cm
2.5 cm
3 cm
3 cm

Area = _____ cm²

⑧
2 m
7.9 m
1.5 m
9.8 m

Area = _____ m²

Area of a trapezoid
= $\frac{1}{2}$ (a+b)h

⑨
5.5 m
4 m
3 m
3.2 m
4.9 m
7.5 m

Area = _____ m²

⑩
7.8 cm
3.7 cm
2 cm
4.2 cm
3 cm
7.5 cm
3.5 cm

Area = _____ cm²

Find the area of the shaded parts.

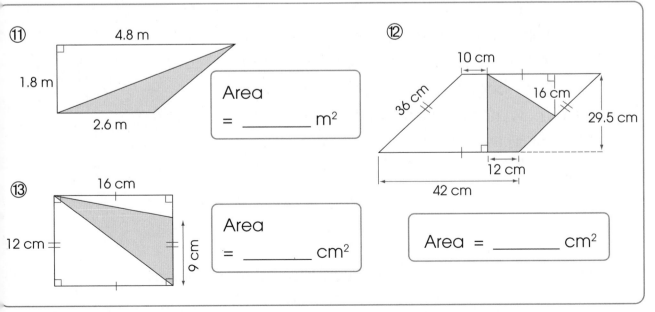

⑪
4.8 m
1.8 m
2.6 m

Area
= _____ m²

⑫
10 cm
36 cm
16 cm
29.5 cm
12 cm
42 cm

⑬
16 cm
12 cm
9 cm

Area
= _____ cm²

Area = _____ cm²

ISBN: 978-1-897164-21-1

Fill in the blanks.

⑭ 14 m² = _____ cm²

⑮ 16.5 m² = _____ cm²

⑯ 8.45 m² = _____ cm²

⑰ 940 m² = _____ cm²

To change from m² to cm², multiply by 10 000, e.g.

42m² = 42×10 000cm²
= 420 000cm²

To change from cm² to m², divide by 10 000, e.g.

45 000cm² = $\frac{45\ 000}{10\ 000}$ m²
= 4.5m²

⑱ 1 260 000 cm² = _____ m²   ⑲ 47 320 cm² = _____ m²

Help Tony solve the problems.

⑳ Tony's bedroom is 4 m long and 3.8 m wide. Carpeting the floor costs $49 per m². How much will it cost to carpet the entire floor?

It costs $_____ to carpet the entire floor.

㉑ The walls of Tony's bedroom are 2.5 m high. Two of the walls are 4 m long and the other two 3.8 m long. If a can of paint can cover 24 m², how many cans of paint will Tony need to paint all the walls?

Tony needs _____ cans of paint to paint all the walls.

ISBN: 978-1-897164-21-1

Tick ✔ the best units for measuring the volume of Tony's boxes.

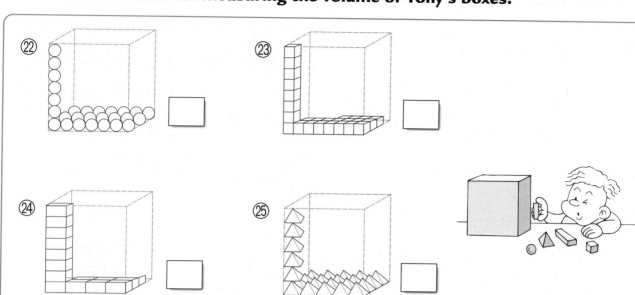

㉒ [ ]

㉓ [ ]

㉔ [ ]

㉕ [ ]

Help Tony find the volume of his blocks.

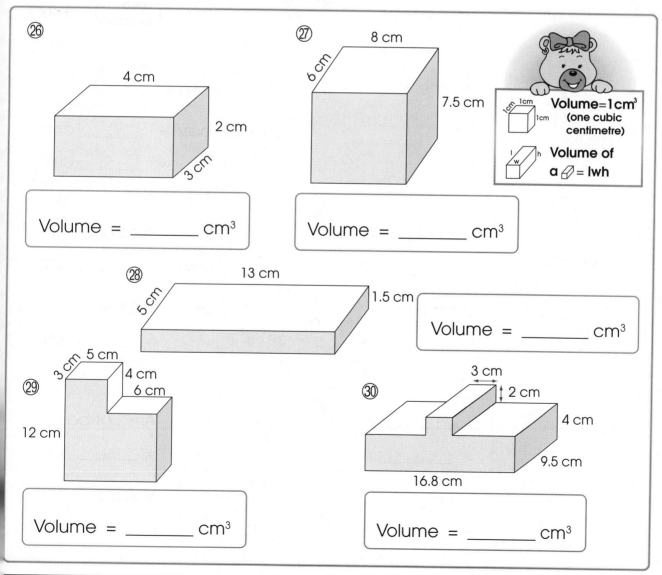

㉖ 4 cm, 2 cm, 3 cm

Volume = _____ cm³

㉗ 8 cm, 6 cm, 7.5 cm

Volume = _____ cm³

Volume = 1cm³
(one cubic centimetre)

Volume of
a ▱ = lwh

㉘ 13 cm, 5 cm, 1.5 cm

Volume = _____ cm³

㉙ 3 cm, 5 cm, 4 cm, 6 cm, 12 cm

Volume = _____ cm³

㉚ 3 cm, 2 cm, 4 cm, 9.5 cm, 16.8 cm

Volume = _____ cm³

ISBN: 978-1-897164-21-1

Help Tony find the area of the base and volume of each block.

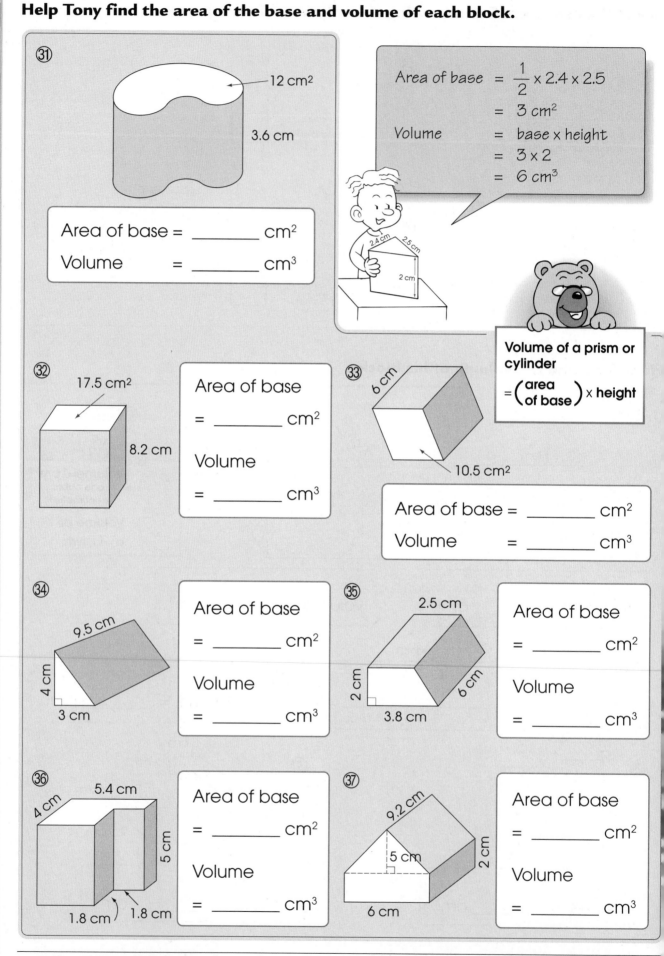

③①

12 cm²

3.6 cm

Area of base = _____ cm²

Volume   = _____ cm³

Area of base = $\frac{1}{2}$ × 2.4 × 2.5

= 3 cm²

Volume = base × height

= 3 × 2

= 6 cm³

2.4 cm   2.5 cm

2 cm

Volume of a prism or cylinder

= $\left(\begin{array}{c}\text{area} \\ \text{of base}\end{array}\right)$ × **height**

③②

17.5 cm²

8.2 cm

Area of base

= _____ cm²

Volume

= _____ cm³

③③

6 cm

10.5 cm²

Area of base = _____ cm²

Volume   = _____ cm³

③④

9.5 cm

4 cm

3 cm

Area of base

= _____ cm²

Volume

= _____ cm³

③⑤

2.5 cm

2 cm

6 cm

3.8 cm

Area of base

= _____ cm²

Volume

= _____ cm³

③⑥

5.4 cm

4 cm

5 cm

1.8 cm   1.8 cm

Area of base

= _____ cm²

Volume

= _____ cm³

③⑦

9.2 cm

5 cm

2 cm

6 cm

Area of base

= _____ cm²

Volume

= _____ cm³

ISBN: 978-1-897164-21-1

Fill in the blanks.

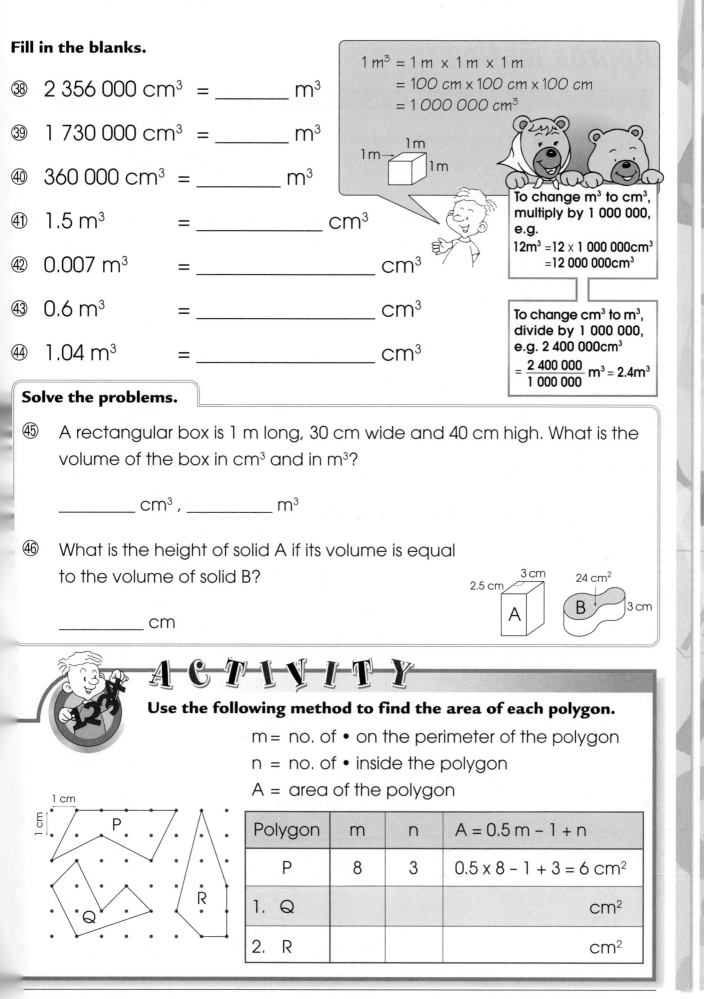

㊳ 2 356 000 cm³ = _____ m³

㊴ 1 730 000 cm³ = _____ m³

㊵ 360 000 cm³ = _____ m³

㊶ 1.5 m³ = _____ cm³

㊷ 0.007 m³ = _____ cm³

㊸ 0.6 m³ = _____ cm³

㊹ 1.04 m³ = _____ cm³

1 m³ = 1 m x 1 m x 1 m
= 100 cm x 100 cm x 100 cm
= 1 000 000 cm³

To change m³ to cm³, multiply by 1 000 000, e.g.
12m³ = 12 x 1 000 000cm³
= 12 000 000cm³

To change cm³ to m³, divide by 1 000 000, e.g. 2 400 000cm³
$= \dfrac{2\ 400\ 000}{1\ 000\ 000}$ m³ = 2.4m³

Solve the problems.

㊺ A rectangular box is 1 m long, 30 cm wide and 40 cm high. What is the volume of the box in cm³ and in m³?

_____ cm³ , _____ m³

㊻ What is the height of solid A if its volume is equal to the volume of solid B?

_____ cm

3 cm
2.5 cm
A

24 cm²
B
3 cm

ACTIVITY

Use the following method to find the area of each polygon.

m = no. of • on the perimeter of the polygon

n = no. of • inside the polygon

A = area of the polygon

Polygon	m	n	A = 0.5 m – 1 + n
P	8	3	0.5 x 8 – 1 + 3 = 6 cm²
1. Q			cm²
2. R			cm²

1 cm
1 cm
P
Q
R

6 Approximation

Tell which of the following show an exact value or an approximate value. Write E for exact and A for approximate.

① The weight of the book is about 1.2 kg. ☐

② The book has 290 pages. ☐

③ The book is about 2.8 cm thick. ☐

④ The cost of the book is $17.25. ☐

⑤ The area of the book is about 602 cm². ☐

Follow Tony's method to find the approximation for the area of each shape.

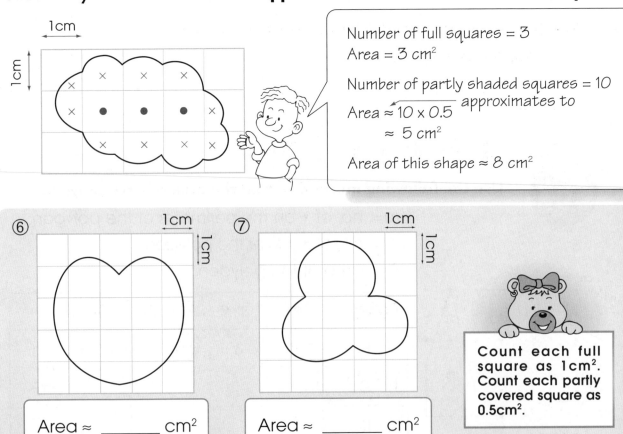

1cm

1cm

Number of full squares = 3
Area = 3 cm²

Number of partly shaded squares = 10

Area ≈ 10 x 0.5 approximates to

≈ 5 cm²

Area of this shape ≈ 8 cm²

⑥

1cm

1cm

Area ≈ _____ cm²

⑦

1cm

1cm

Area ≈ _____ cm²

Count each full square as 1cm². Count each partly covered square as 0.5cm².

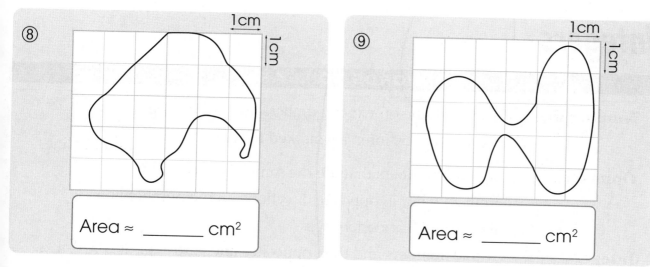

⑧ Area ≈ _____ cm²

⑨ Area ≈ _____ cm²

Help Dave complete the table.

The length of AB to the nearest cm : AB ≈ 5 cm

The length of AB to the nearest 0.5 cm : AB ≈ 4.5 cm

Line	Length to the nearest cm	Length to the nearest 0.5 cm
⑩ MN		
⑪ PQ		
⑫ RS		
⑬ MS		

ACTIVITY

Estimate the number of marbles that can be put into the box. Then find the exact number.

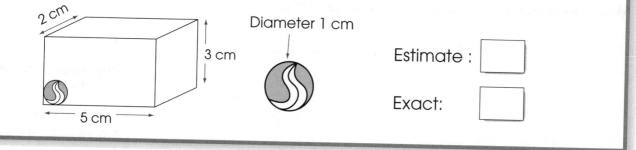

2 cm

3 cm

5 cm

Diameter 1 cm

Estimate : ☐

Exact: ☐

ISBN: 978-1-897164-21-1

7 Integers

Number line	-	a straight line on which numbers are represented by intervals marked to scale
Opposites	-	the two numbers that are the same distance from zero but in opposite directions on the number line eg. –5 and 5 are opposites.
Integers	-	whole numbers and their opposites and zero eg. 20, –5, 0, 4, ...

eg.
 –1 0 1 2 <br> number line

Follow Dave's method to write the numbers as integers.

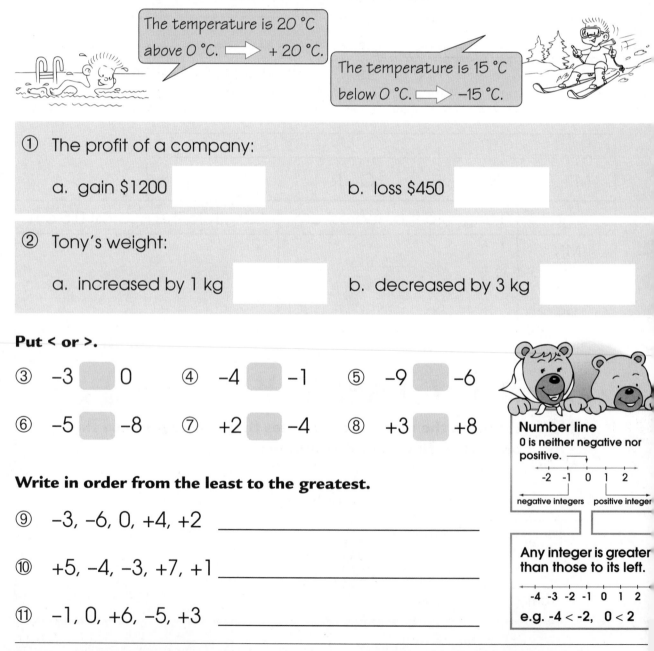

The temperature is 20 °C above 0 °C. ⟹ + 20 °C.

The temperature is 15 °C below 0 °C. ⟹ –15 °C.

① The profit of a company:

   a. gain $1200   b. loss $450

② Tony's weight:

   a. increased by 1 kg   b. decreased by 3 kg

Put < or >.

③   –3 ☐ 0   ④   –4 ☐ –1   ⑤   –9 ☐ –6

⑥   –5 ☐ –8   ⑦   +2 ☐ –4   ⑧   +3 ☐ +8

Number line
0 is neither negative nor positive.

-2 -1 0 1 2

negative integers   positive integer

Write in order from the least to the greatest.

⑨   –3, –6, 0, +4, +2   _____

⑩   +5, –4, –3, +7, +1   _____

⑪   –1, 0, +6, –5, +3   _____

Any integer is greater than those to its left.

-4 -3 -2 -1 0 1 2

e.g. -4 < -2,   0 < 2

ISBN: 978-1-897164-21-1

Follow Dave's method to write an addition sentence for each diagram.

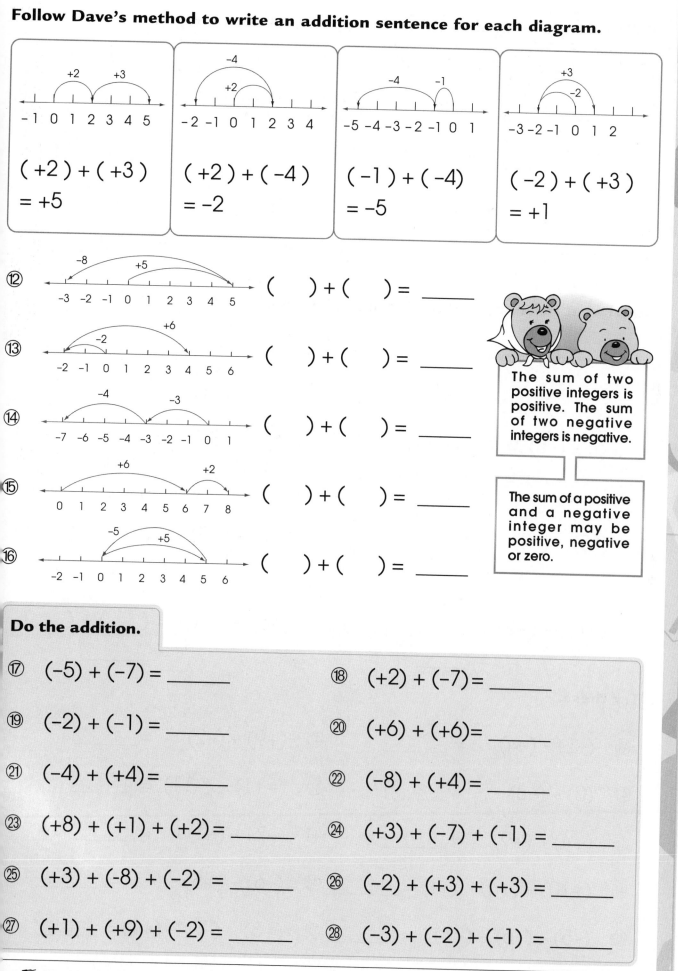

(+2) + (+3)
= +5

(+2) + (–4)
= –2

(–1) + (–4)
= –5

(–2) + (+3)
= +1

⑫ (   ) + (   ) = _____

⑬ (   ) + (   ) = _____

⑭ (   ) + (   ) = _____

⑮ (   ) + (   ) = _____

⑯ (   ) + (   ) = _____

The sum of two positive integers is positive. The sum of two negative integers is negative.

The sum of a positive and a negative integer may be positive, negative or zero.

Do the addition.

⑰ (–5) + (–7) = _____

⑱ (+2) + (–7) = _____

⑲ (–2) + (–1) = _____

⑳ (+6) + (+6) = _____

㉑ (–4) + (+4) = _____

㉒ (–8) + (+4) = _____

㉓ (+8) + (+1) + (+2) = _____

㉔ (+3) + (–7) + (–1) = _____

㉕ (+3) + (-8) + (–2) = _____

㉖ (–2) + (+3) + (+3) = _____

㉗ (+1) + (+9) + (–2) = _____

㉘ (–3) + (–2) + (–1) = _____

ISBN: 978-1-897164-21-1

Follow Tony's method to write the opposite of each integer.

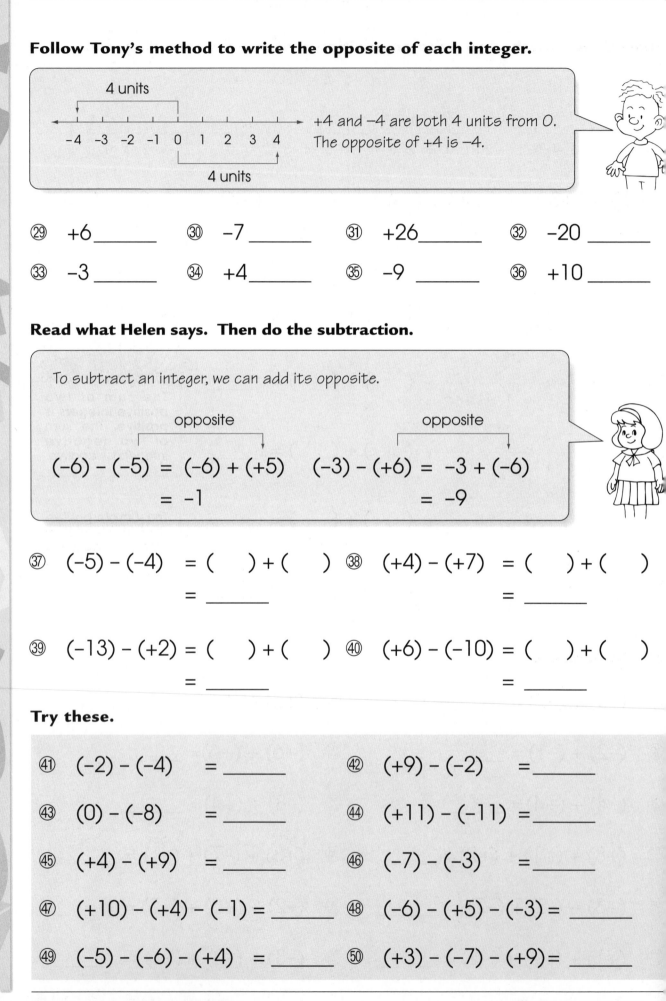

4 units

-4 -3 -2 -1 0 1 2 3 4

4 units

+4 and −4 are both 4 units from 0. The opposite of +4 is −4.

㉙　+6 _____　　㉚　−7 _____　　㉛　+26 _____　　㉜　−20 _____

㉝　−3 _____　　㉞　+4 _____　　㉟　−9 _____　　㊱　+10 _____

Read what Helen says. Then do the subtraction.

To subtract an integer, we can add its opposite.

opposite

opposite

$(-6) - (-5) = (-6) + (+5)$　　$(-3) - (+6) = -3 + (-6)$
$= -1$　　　　　　　　　　　　$= -9$

㊲　$(-5) - (-4) = (\quad) + (\quad)$　　㊳　$(+4) - (+7) = (\quad) + (\quad)$
　　　　　　$= \_\_\_\_\_$　　　　　　　　　　　　$= \_\_\_\_\_$

㊴　$(-13) - (+2) = (\quad) + (\quad)$　　㊵　$(+6) - (-10) = (\quad) + (\quad)$
　　　　　　　$= \_\_\_\_\_$　　　　　　　　　　　$= \_\_\_\_\_$

Try these.

㊶　$(-2) - (-4) = \_\_\_\_\_$　　　　㊷　$(+9) - (-2) = \_\_\_\_\_$

㊸　$(0) - (-8) = \_\_\_\_\_$　　　　㊹　$(+11) - (-11) = \_\_\_\_\_$

㊺　$(+4) - (+9) = \_\_\_\_\_$　　　　㊻　$(-7) - (-3) = \_\_\_\_\_$

㊼　$(+10) - (+4) - (-1) = \_\_\_\_\_$　　㊽　$(-6) - (+5) - (-3) = \_\_\_\_\_$

㊾　$(-5) - (-6) - (+4) = \_\_\_\_\_$　　㊿　$(+3) - (-7) - (+9) = \_\_\_\_\_$

ISBN: 978-1-897164-21-1

Read Tony's note and solve the problems. Then write the letters to find what Tony says.

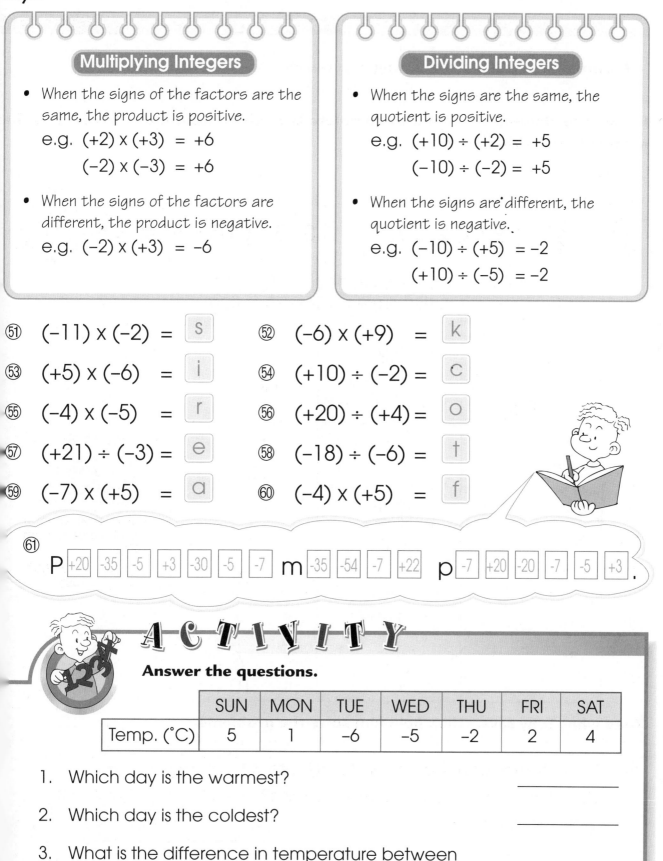

Multiplying Integers

- When the signs of the factors are the same, the product is positive.

  e.g. $(+2) \times (+3) = +6$

  $(-2) \times (-3) = +6$

- When the signs of the factors are different, the product is negative.

  e.g. $(-2) \times (+3) = -6$

Dividing Integers

- When the signs are the same, the quotient is positive.

  e.g. $(+10) \div (+2) = +5$

  $(-10) \div (-2) = +5$

- When the signs are different, the quotient is negative.

  e.g. $(-10) \div (+5) = -2$

  $(+10) \div (-5) = -2$

(51) $(-11) \times (-2) =$ ____ s

(52) $(-6) \times (+9) =$ ____ k

(53) $(+5) \times (-6) =$ ____ i

(54) $(+10) \div (-2) =$ ____ c

(55) $(-4) \times (-5) =$ ____ r

(56) $(+20) \div (+4) =$ ____ o

(57) $(+21) \div (-3) =$ ____ e

(58) $(-18) \div (-6) =$ ____ t

(59) $(-7) \times (+5) =$ ____ a

(60) $(-4) \times (+5) =$ ____ f

(61) P [+20] [-35] [-5] [+3] [-30] [-5] [-7] m [-35] [-54] [-7] [+22] p [-7] [+20] [-20] [-7] [-5] [+3] .

ACTIVITY

Answer the questions.

	SUN	MON	TUE	WED	THU	FRI	SAT
Temp. (°C)	5	1	-6	-5	-2	2	4

1. Which day is the warmest? ________

2. Which day is the coldest? ________

3. What is the difference in temperature between Wednesday and Saturday? ________ °C

ISBN: 978-1-897164-21-1

8 Fractions, Decimals, and Percents

Formula - a general rule showing how variables are related to one another

Follow the children's method to complete the table. Round the answers to the nearest hundredth, if necessary.

$$\frac{3}{4} = \frac{3}{4} \times 100\% = 75\%$$

$$0.05 = 0.05 \times 100\% = 5\%$$

	Percent	Decimal	Fraction
①		0.2	
②	28%		
③			$\frac{9}{20}$
④		0.65	
⑤			$\frac{2}{3}$
⑥		1.4	
⑦			$2\frac{7}{10}$

Any number greater than 1 is greater than 100%.
e.g. $1.5 = 1.5 \times 100\%$
$= 150\%$

Read what Tony says. Then help him solve the problems.

⑧ Tony ate 10% of 20 apples and Dave ate 12% of 25 apples. Who ate more apples? How many more?

10% of 20
= 10% × 20 = 0.1 × 20 = 2
I ate 2 apples.

_____ate _____ more apple(s) than _____.

COMPLETE MATHSMART (GRADE 7)   ISBN: 978-1-897164-21-1

⑨ Helen spent 23.5% of $40 and Elaine spent $\frac{1}{8}$ of $74. Who spent more money? How much more?

_____ spent $_____ more than _____ .

⑩ Tony answered 84% of the 50 questions on the history test correctly. He answered 0.75 of the 60 questions on the math test correctly. On which test did he get more correct answers? How many more?

Tony got _____ more correct answers on the _____ test

than on the _____ test.

Use the rules below to find the answers.

⑪ $(23 + 2) \div 5 + 6$ = _____

⑫ $24 \div (6 + 2) \times 5$ = _____

⑬ $(42 - 2) \div (4 + 6)$ = _____

⑭ $0.5 \times (18 + 2) \div \frac{1}{2}$ = _____

⑮ $(0.6 \times 10)^2 \div (11 + 1)$ = _____

⑯ $18 - (0.2 + 40\%) \times 8$ = _____

⑰ $0.9 + (4 + 2)^2 \div 2$ = _____

⑱ $0.5 \times (1.6 - 0.2) \div 4$ = _____

⑲ $650\% - (0.7 + 0.5)^2 \times 4 =$ _____

Order of Operations

1st   Do all operations inside the brackets.

2nd   Do exponents.

3rd   Do multiplication and division from left to right.

4th   Do addition and subtraction from left to right.

ISBN: 978-1-897164-21-1

Tick ✔ the correct answers and grade the children's test papers.

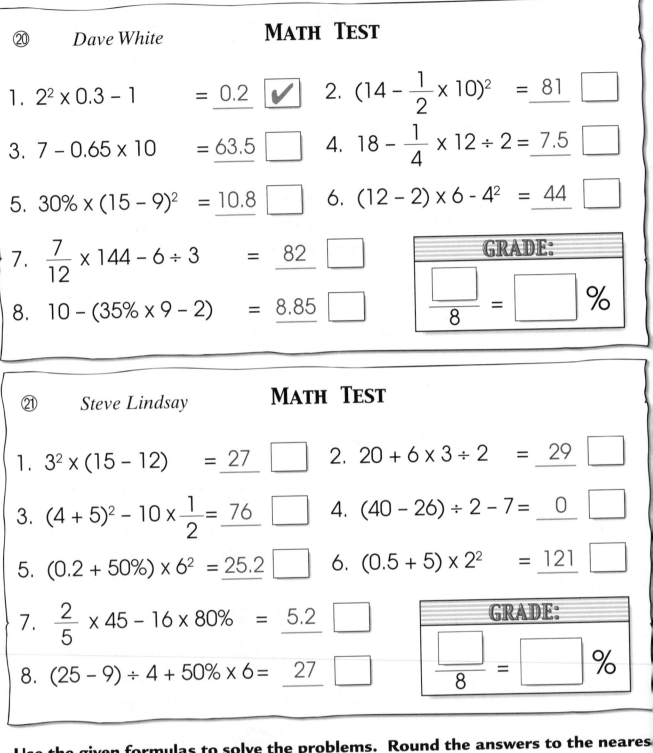

⑳   *Dave White*   MATH TEST

1. $2^2 \times 0.3 - 1$ = <u>0.2</u> ✔   2. $(14 - \frac{1}{2} \times 10)^2$ = <u>81</u> ☐

3. $7 - 0.65 \times 10$ = <u>63.5</u> ☐   4. $18 - \frac{1}{4} \times 12 \div 2$ = <u>7.5</u> ☐

5. $30\% \times (15 - 9)^2$ = <u>10.8</u> ☐   6. $(12 - 2) \times 6 - 4^2$ = <u>44</u> ☐

7. $\frac{7}{12} \times 144 - 6 \div 3$ = <u>82</u> ☐

8. $10 - (35\% \times 9 - 2)$ = <u>8.85</u> ☐

GRADE:
$\frac{\boxed{}}{8} = \boxed{}\%$

㉑   *Steve Lindsay*   MATH TEST

1. $3^2 \times (15 - 12)$ = <u>27</u> ☐   2. $20 + 6 \times 3 \div 2$ = <u>29</u> ☐

3. $(4 + 5)^2 - 10 \times \frac{1}{2}$ = <u>76</u> ☐   4. $(40 - 26) \div 2 - 7$ = <u>0</u> ☐

5. $(0.2 + 50\%) \times 6^2$ = <u>25.2</u> ☐   6. $(0.5 + 5) \times 2^2$ = <u>121</u> ☐

7. $\frac{2}{5} \times 45 - 16 \times 80\%$ = <u>5.2</u> ☐

8. $(25 - 9) \div 4 + 50\% \times 6$ = <u>27</u> ☐

GRADE:
$\frac{\boxed{}}{8} = \boxed{}\%$

Use the given formulas to solve the problems. Round the answers to the neares hundredth, if necessary.

㉒   Uncle Paul's wages:

Wages = \$500 + \$10.5 × (number of working hours)

a.   If he works $4\frac{1}{2}$ hours, how much will he get?   \$_____

b.   If he works 8 hours, how much will he get?   \$_____

ISBN: 978-1-897164-21-1

㉓ **Taxi fare:**

Taxi fare = $2.50 + $0.75 x (distance travelled in km)

a. If Tony travelled 8.7 km, how much would the fare be? $_____

b. If Tony travelled 10.25 km, how much would the fare be? $_____

㉔ **The weight of a box of candies:**

Weight = 400 g – 15 g x (number of candies eaten)

a. If Dave ate 5 candies, how heavy would the box be?   _____ g

b. If Dave ate 12 candies, how heavy would the box be?   _____ g

㉕ **The money that Helen has:**

Helen's money = (Dave's money + Steve's money) x 40%

a. If Dave has $25 and Steve has $35,

how much does Helen have? $_____

b. If Dave has $70.50 and Steve has $16,

how much does Helen have? $_____

A C T I V I T Y

Put brackets and operation signs to these numbers to produce the right answers.

2 x (3 + 4) – 5 + 1 = 10

1. 5 ▢ 2 ▢ 4 ▢ 3 ▢ 1 = 10

2. 3 ▢ 2 ▢ 1 ▢ 5 ▢ 4 = 10

3. 5 ▢ 2 ▢ 1 ▢ 4 ▢ 3 = 10

4. ((4 ▢ 2) ▢ 5) ▢ 3 ▢ 1 = 10

ISBN: 978-1-897164-21-1

Midway Test

Write each product as a power and state the base and exponent of each power. (6 marks)

① $8 \times 8 \times 8 \times 8 \times 8 \times 8 \times 8 \times 8 \times 8 \times 8 \times 8 \times 8 =$ _____

Exponent = _____   Base = _____

② Four to the fifth power = _____

Exponent = _____   Base = _____

Find the prime factors of each number and write as a power. Then find the square root of each number. (8 marks)

③ $81 =$ _____
$\sqrt{81} =$ _____

④ $16 =$ _____
$\sqrt{16} =$ _____

⑤ $64 =$ _____
$\sqrt{64} =$ _____

⑥ $400 =$ _____ X _____
$\sqrt{400} =$ _____

⑦ $196 =$ _____ X _____
$\sqrt{196} =$ _____

⑧ $484 =$ _____ X _____
$\sqrt{484} =$ _____

Solve the equations. (8 marks)

⑨ $5 + y = 9$ $y =$ _____

⑩ $m - 7 = 14$ $m =$ _____

⑪ $p - 2.5 = 6.3$ $p =$ _____

⑫ $q + 1.4 = 2.7$ $q =$ _____

⑬ $0.2w = 5$ $w =$ _____

⑭ $t \div 6 = 0.9$ $t =$ _____

⑮ $\dfrac{1}{4} k = 0.3$ $k =$ _____

⑯ $z \div 0.5 = 4$ $z =$ _____

ISBN: 978-1-897164-21-1

Write the answers in simplest form. (8 marks)

⑰ $2\dfrac{1}{3} \times \dfrac{4}{7} =$ _____

⑱ $5\dfrac{1}{6} \times 3 =$ _____

⑲ $1\dfrac{2}{5} \div \dfrac{7}{15} =$ _____

⑳ $2\dfrac{2}{5} \div 8 =$ _____

㉑ $\dfrac{2}{3} \times \left(\dfrac{2}{5} + \dfrac{4}{5} \right) =$ _____

㉒ $\left(\dfrac{1}{6} + 1\dfrac{1}{2} \right) \div \dfrac{5}{6} =$ _____

㉓ $\dfrac{5}{16} \div \left(2\dfrac{1}{8} - 1\dfrac{1}{2} \right) =$ _____

㉔ $\dfrac{4}{9} \times \left(2\dfrac{1}{12} - 1\dfrac{1}{3} \right) =$ _____

Solve the problems. (6 marks)

㉕ Each cup can hold $\dfrac{1}{6}$ L of juice. How many cups are needed for $1\dfrac{1}{3}$ L of juice?

_____ cups

㉖ A mug can hold $\dfrac{7}{8}$ L of water. A glass can only hold $\dfrac{2}{3}$ as much. How much water can the glass hold?

_____ L

㉗ A car travels $69\dfrac{1}{3}$ km an hour. How far can it travel in $1\dfrac{1}{2}$ hours?

_____ km

Find the percent of the shaded parts in each shape. Round to the nearest hundredth, if necessary. (2 marks)

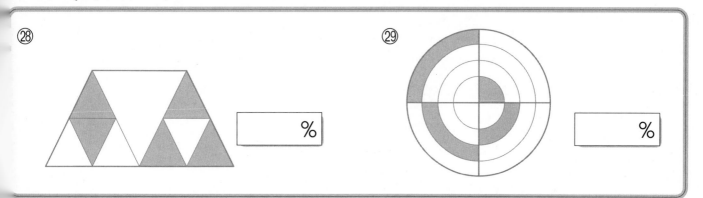

㉘ _____ %

㉙ _____ %

ISBN: 978-1-897164-21-1

Complete the tables. (14 marks)

	Regular Price	Discount Rate	Amount of Discount	Sale Price
㉚	$40	25%		
㉛	$75	30%		
㉜			$25.20	$100.80
㉝	$89.20		$13.38	

	Selling Price	Tax Rate	Amount of Tax	Total Cost
㉞	$36	12%		
㉟	$15.50		$1.24	
㊱			$11.34	$86.94

Find the perimeter and area of each shape. (8 marks)

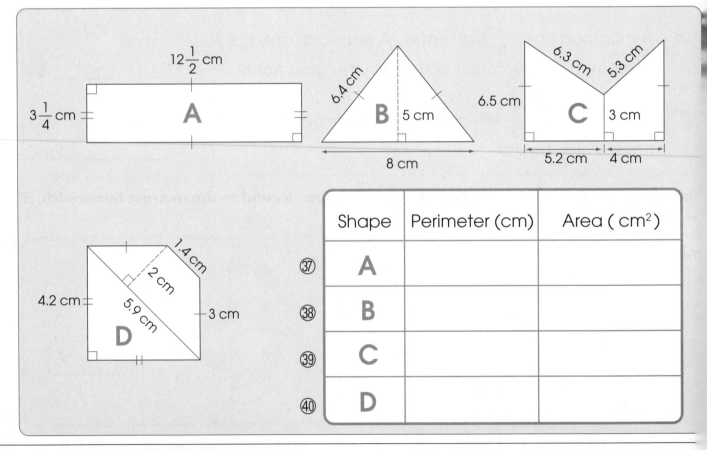

Shape	Perimeter (cm)	Area (cm²)
㊲ A		
㊳ B		
㊴ C		
㊵ D		

ISBN: 978-1-897164-21-1

Find the volume of each solid. (8 marks)

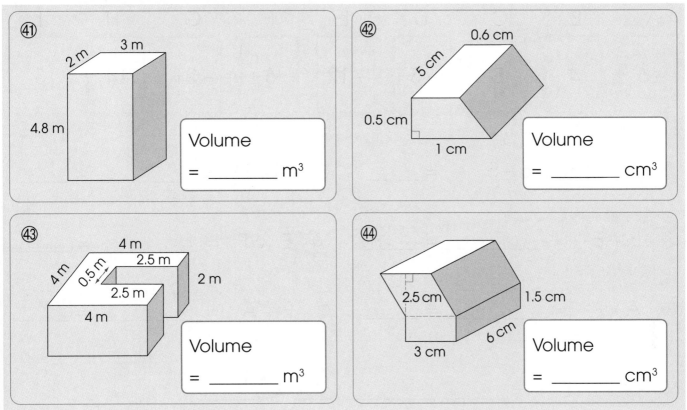

㊶

2 m 3 m

4.8 m

Volume

= _____ m³

㊷

0.6 cm

5 cm

0.5 cm

1 cm

Volume

= _____ cm³

㊸

4 m

2.5 m

4 m 0.5 m

2.5 m

4 m

2 m

Volume

= _____ m³

㊹

2.5 cm

1.5 cm

3 cm 6 cm

Volume

= _____ cm³

Fill in the blanks. (4 marks)

㊺ 5 200 cm² = _____ m²	㊻ 1.85 m² = _____ cm²
㊼ 4 560 000 cm³ = _____ m³	㊽ 0.85 m³ = _____ cm³

Complete the table. (6 marks)

| | | A | | B | | C | | D |

Line	Length to the nearest cm	Length to the nearest 0.5 cm
㊾ AB		
50 AC		
51 AD		

ISBN: 978-1-897164-21-1

Midway Test

Use the number cards below to solve the problems. (8 marks)

A	B	C	D	E	F	G	H	I
-6	4	5	-3	-12	6	-21	14	3

52 $A - C = -6 - 5 = \underline{\hspace{1.5cm}}$

53 $G \div D = \underline{\hspace{3cm}} = \underline{\hspace{1.5cm}}$

54 $B + E = \underline{\hspace{3cm}} = \underline{\hspace{1.5cm}}$

55 $E \times F = \underline{\hspace{3cm}} = \underline{\hspace{1.5cm}}$

56 $A \times I = \underline{\hspace{3cm}} = \underline{\hspace{1.5cm}}$

57 $E \div A = \underline{\hspace{3cm}} = \underline{\hspace{1.5cm}}$

58 $G + I - E = \underline{\hspace{5cm}} = \underline{\hspace{2cm}}$

59 $C - (A + H) = \underline{\hspace{5cm}} = \underline{\hspace{2cm}}$

Write an integer equation for each problem. Then solve the problems. (4 marks)

60 Two integers have a sum of +5. One of the integers is +12. What is the other integer?

The other integer is _____ .

61 The temperature dropped 11 °C from a high of 7 °C. What was the temperature after the drop?

The temperature was _____ °C.

ISBN: 978-1-897164-21-1

Complete the table. (4 marks)

	Decimal	Fraction	Percent
⑥2	1.6		
⑥3			45%

Solve the problems. (4 marks)

⑥4 Dave saved 38.5% of $70 and Steve saved 46.3% of $50. Who saved more money? How much more?

_____ saved $_____ more than _____ .

⑥5 Uncle Fred sold 85% of the 80 cakes he had.

a. How many cakes did he sell? _____ cakes

b. If he had sold 4 more cakes, what percent would he have sold altogether? _____%

Use the given formula to answer the questions. (2 marks)

⑥6 Steve's savings:

Savings = $78.50 – $6.50 x number of days

SCORE

100

a. How much will he have after 7 days? $_____

b. How much will he have after 12 days? $_____

⑨ Coordinates

Coordinate plane - a plane divided into 4 quadrants to represent ordered pairs of numbers as points

Linear equation - an equation represented by a straight line on a coordinate plane

Solution - the value of the variable(s) that makes the equation true

Write the ordered pair to show the location of each labelled point on the grid.

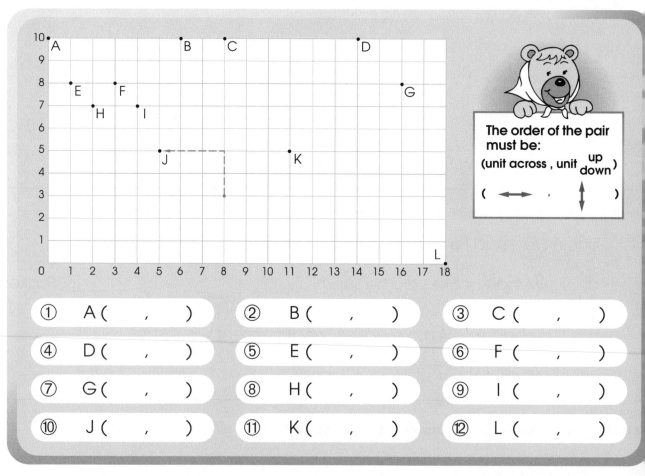

The order of the pair must be:

(unit across , unit $^{up}_{down}$)

(⟷ , ↕)

①	A (,)	②	B (,)	③	C (,)
④	D (,)	⑤	E (,)	⑥	F (,)
⑦	G (,)	⑧	H (,)	⑨	I (,)
⑩	J (,)	⑪	K (,)	⑫	L (,)

Follow Tony's method to locate each child on the grid above and find which letter they will reach.

I am at (8, 3). If I move 2 units up and 3 units left, I will reach point "J".

⑬ Dave is at (3, 2). If he moves 5 units up and 1 unit right, he will reach point _____ .

⑭ Helen is at (0, 10). If she moves 8 units right, she will reach point _____ .

⑮ Elaine is at (9, 7). If she moves 7 units right and 1 unit up, she will reach point _____ .

⑯ Steve is at (16, 7). If he moves 5 units left and 2 units down, he will reach point _____ .

Use the coordinate plane below to answer the questions.

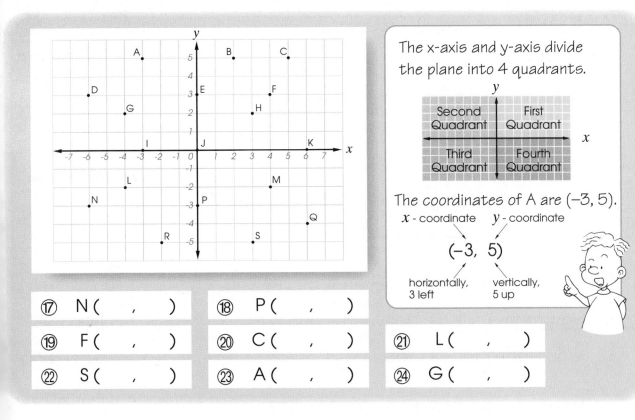

The x-axis and y-axis divide the plane into 4 quadrants.

The coordinates of A are (−3, 5).

x - coordinate   y - coordinate

(−3, 5)

horizontally, 3 left   vertically, 5 up

⑰ N (,)   ⑱ P (,)

⑲ F (,)   ⑳ C (,)   ㉑ L (,)

㉒ S (,)   ㉓ A (,)   ㉔ G (,)

㉕ Which points are in the first quadrant? _____

㉖ Which points are in the second quadrant? _____

㉗ Which points are in the third quadrant? _____

㉘ Which points are in the fourth quadrant? _____

㉙ Which points are on the x-axis? _____

㉚ Which points are on the y-axis? _____

Look at Tony's coordinate plane and answer the questions.

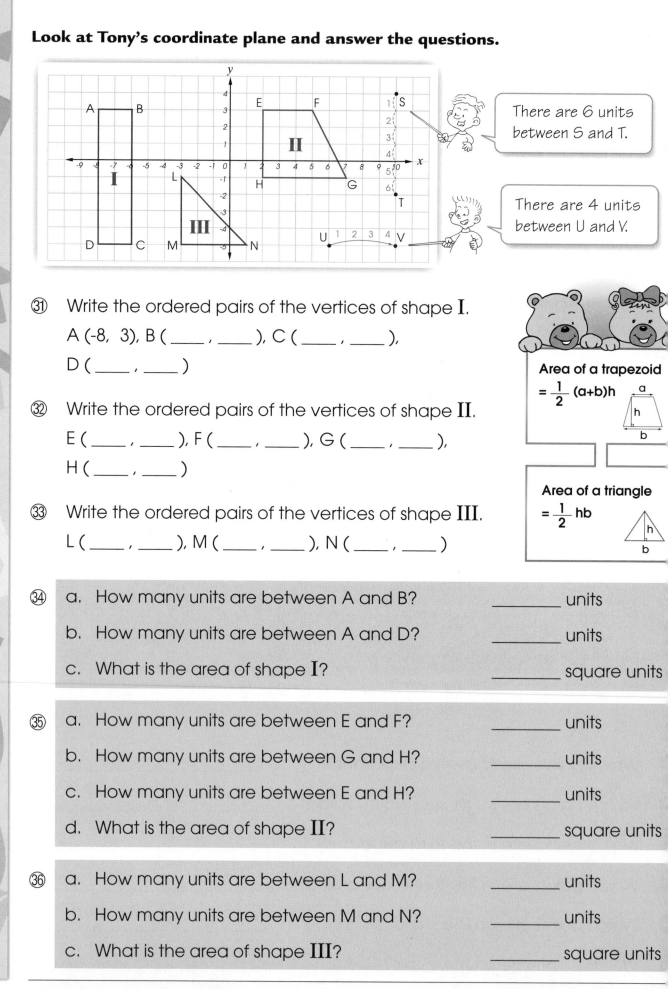

There are 6 units between S and T.

There are 4 units between U and V.

㉛ Write the ordered pairs of the vertices of shape **I**.

A (-8, 3), B (____ , ____), C (____ , ____),

D (____ , ____)

㉜ Write the ordered pairs of the vertices of shape **II**.

E (____ , ____), F (____ , ____), G (____ , ____),

H (____ , ____)

Area of a trapezoid
$= \dfrac{1}{2} (a+b)h$

㉝ Write the ordered pairs of the vertices of shape **III**.

L (____ , ____), M (____ , ____), N (____ , ____)

Area of a triangle
$= \dfrac{1}{2} hb$

㉞ a. How many units are between A and B? _____ units

b. How many units are between A and D? _____ units

c. What is the area of shape **I**? _____ square units

㉟ a. How many units are between E and F? _____ units

b. How many units are between G and H? _____ units

c. How many units are between E and H? _____ units

d. What is the area of shape **II**? _____ square units

㊱ a. How many units are between L and M? _____ units

b. How many units are between M and N? _____ units

c. What is the area of shape **III**? _____ square units

ISBN: 978-1-897164-21-1

Follow Tony's method to complete the tables and draw the graphs. Then use the graphs to answer the questions.

$y = x + 2$

x	-1	0	2
y	1	2	4
	$y = -1 + 2$ $= 1$	$y = 0 + 2$ $= 2$	$y = 2 + 2$ $= 4$
(x, y)	(-1, 1)	(0, 2)	(2, 4)

From the graph you can see that when $x = 1$, $y = 3$.

⑰ $y = x - 1$

a.

x	0	3	5
y			
(x, y)			

b. When $x = 4$, $y =$ _____ .

c. When $y = 1$, $x =$ _____ .

⑱ $y = 2x - 1$

a.

x	-1	1	3
y			
(x, y)			

b. When $x = 2$, $y =$ _____ .

c. When $y = -1$, $x =$ _____ .

ACTIVITY

Look at the graph and tick ✔ the right boxes.

1. Equation A is ☐ $y = x + 1$ ☐ $y = x + 2$

2. Equation B is ☐ $x = 1$ ☐ $y = 1$

3. The point of intersection of equations A and B is
☐ (1, 3) ☐ (3, 1)

1₀ *More about Algebraic Expressions*

Term	-	a mathematical expression that is either a number or the product of a number and one or more variables
Variable	-	an unknown value represented by a letter
Literal part	-	the variable part of a term

e.g. $3x^2 + 5$

Term: $3x^2, +5$ Literal part of $3x^2$: x^2

Evaluate	-	find out an idea of the amount or value of something
Formula	-	a general rule showing how variables are related to one another

Read what the children say. Then write in algebraic expressions.

①  A number 25 less than the square of y. _____

②  The sum of 25 and y, divided by 5. _____

③  The sum of 25 and the square of y. _____

④  The square of the difference between y and 25. _____

⑤ The sum of 25 and y, multiplied by 2. _____

⑥ Subtract 25 from the product of y and 25. _____

Evaluate for $y = 6$.

⑦ $2y + 12 =$ _____

⑧ $y^2 + y =$ _____

⑨ $9(y - 2) =$ _____

⑩ $(10.5 - y) \div 5 =$ _____

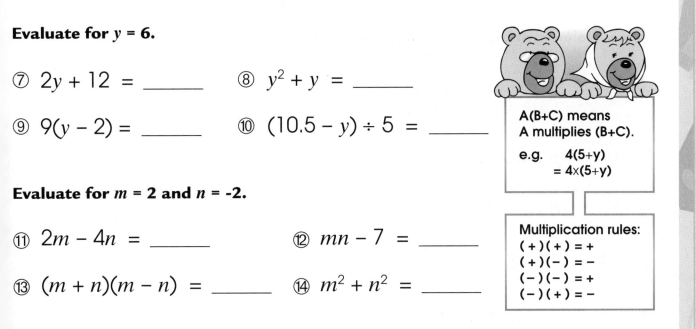

A(B+C) means
A multiplies (B+C).

e.g.   4(5+y)
       = 4×(5+y)

Evaluate for $m = 2$ and $n = -2$.

⑪ $2m - 4n =$ _____

⑫ $mn - 7 =$ _____

⑬ $(m + n)(m - n) =$ _____

⑭ $m^2 + n^2 =$ _____

Multiplication rules:
(+)(+) = +
(+)(−) = −
(−)(−) = +
(−)(+) = −

Read what Tony says. Then help him complete the table.

Expression   : $2y^2 - 3y + 7$
Terms   : $2y^2$, $-3y$, 7
No. of terms : 3

An algebraic expression can be considered as the sum of the terms.
i.e. $2y^2 + (-3y) + 7$

	Expression	Terms	No. of Terms
⑮	$5x^2 + 3x + 2y - 8$		
⑯	$3a + 2b - 6$		
⑰	$0.5m + 2n - 2p + 9$		
⑱	$8u^3 - 4u^2 + 3$		

Write whether the terms are like or unlike.

⑲ x^2, $3x^2$ and $-0.5x^2$ are _____ terms.

⑳ $-3y$, y^2 and $2y$ are _____ terms.

㉑ $4m$, $\frac{3}{4}m$, $0.6m$ and $-m$ are _____ terms.

㉒ $10xy$, $-2xy^2$, $\frac{9}{5}xy$ and xy^2 are _____ terms.

Terms with the same literal parts are called like terms. Otherwise, they are unlike terms.

Follow Helen's method to simplify the expressions.

> We can use the distributive law to combine like terms.
> 5x and 2x are like terms; 9 and -4 are like terms.

$5x + 9 + 2x - 4$
$= 5x + 2x + 9 - 4$
$= (5 + 2)x + 5$
$= 7x + 5$

㉓ $6p + 4p + 10 - 6 =$ _____

㉔ $8d - 2d - 2 + 15 =$ _____

㉕ $3k - 12k - 3 - 6 =$ _____

㉖ $\dfrac{3}{4}m + \dfrac{1}{2}m + 4 - 11 \ =$ _____

㉗ $8n^2 - 6n + 7n^2 - 4n + 5 \quad =$ _____

Distributive Law

$a(b + c) = ab + ac$

$mp + mq = m(p + q)$

㉘ $3a + 4b - 9a + 4 - b \quad =$ _____

㉙ $m + 2n - 6n + 8m + 5 \quad =$ _____

㉚ $2pq + 4p + 3p - 6pq - 7p \ =$ _____

㉛ $6x^2y + 5y - x^2y + 3y - 6y \quad =$ _____

Simplify the expressions.

㉜ $3(x + 5) - 6$	㉝ $5y - 2(y + 1)$
$=$ _____ $x +$ _____	$=$ _____
㉞ $6(h - 2) + 2h$	㉟ $-2(m + 7) + 10$
$=$ _____	$=$ _____
㊱ $8n + 4(3 - n)$	㊲ $3(p + 4) + 2(p - 2)$
$=$ _____	$=$ _____

ISBN: 978-1-897164-21-1

Evaluate for $n = 3$.

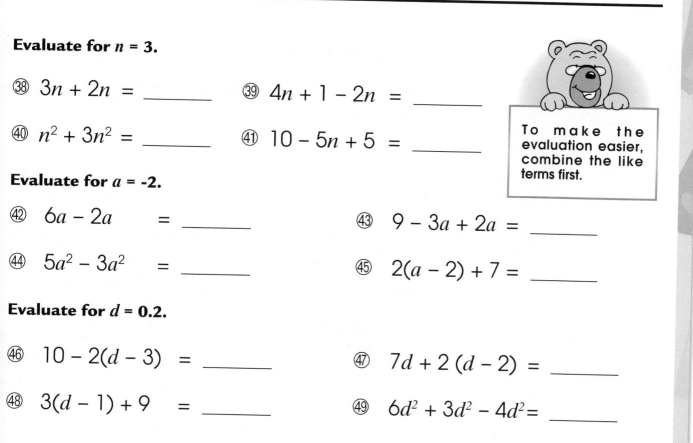

㊳ $3n + 2n = $ _____

㊴ $4n + 1 - 2n = $ _____

㊵ $n^2 + 3n^2 = $ _____

㊶ $10 - 5n + 5 = $ _____

> To make the evaluation easier, combine the like terms first.

Evaluate for $a = -2$.

㊷ $6a - 2a = $ _____

㊸ $9 - 3a + 2a = $ _____

㊹ $5a^2 - 3a^2 = $ _____

㊺ $2(a - 2) + 7 = $ _____

Evaluate for $d = 0.2$.

㊻ $10 - 2(d - 3) = $ _____

㊼ $7d + 2(d - 2) = $ _____

㊽ $3(d - 1) + 9 = $ _____

㊾ $6d^2 + 3d^2 - 4d^2 = $ _____

Read what the children say. Then substitute the values for the letters and calculate.

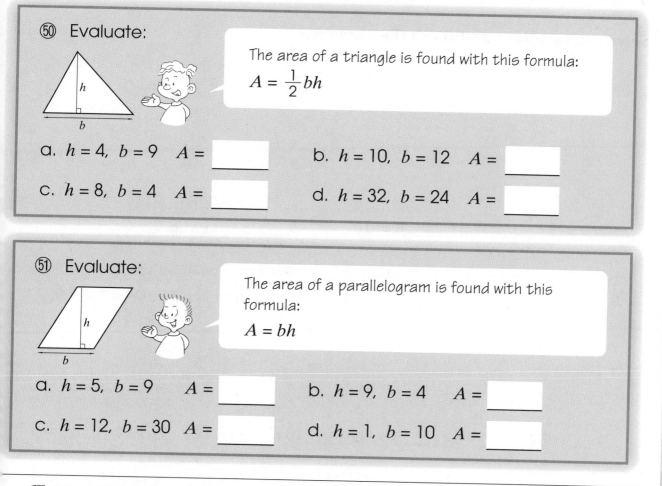

㊿ Evaluate:

The area of a triangle is found with this formula:
$A = \frac{1}{2}bh$

a. $h = 4$, $b = 9$ $A = $ ____

b. $h = 10$, $b = 12$ $A = $ ____

c. $h = 8$, $b = 4$ $A = $ ____

d. $h = 32$, $b = 24$ $A = $ ____

51 Evaluate:

The area of a parallelogram is found with this formula:
$A = bh$

a. $h = 5$, $b = 9$ $A = $ ____

b. $h = 9$, $b = 4$ $A = $ ____

c. $h = 12$, $b = 30$ $A = $ ____

d. $h = 1$, $b = 10$ $A = $ ____

ISBN: 978-1-897164-21-1

52. **Evaluate:**

The volume of a rectangular block is found with this formula:

$$V = lwh$$

a. $l = 3$, $w = 5$, $h = 4$ $V =$ _____

b. $l = 5$, $w = 10$, $h = 2$ $V =$ _____

c. $l = 0.5$, $w = 4$, $h = 7$ $V =$ _____

d. $l = 0.2$, $w = 3$, $h = 0.5$ $V =$ _____

53. **Evaluate:**

Use this formula to find the percent scored on a test, where $r =$ number of right answers and $n =$ number of problems.

$$P = \frac{r}{n} \times 100\%$$

a. $r = 70$, $n = 80$ $P =$ _____

b. $r = 65$, $n = 200$ $P =$ _____

c. $r = 30$, $n = 32$ $P =$ _____

d. $r = 36$, $n = 45$ $P =$ _____

Solve the equations and check the answers.

54. $3y + 5y = 16$

Check: $3(\underline{\quad}) + 5(\underline{\quad}) = 16$

55. $x - 7 + 4 = 6$

Check: $(\underline{\quad}) - 7 + 4 = 6$

56. $8m - 6m = -12$

Check: $8(\underline{\quad}) - 6(\underline{\quad}) = -12$

57. $8a - 12a + 6a = 14$

Check:

$8(\underline{\quad}) - 12(\underline{\quad}) + 6(\underline{\quad}) = 14$

ISBN: 978-1-897164-21-1

Help Uncle Fred complete the equations for his problems. Then find the answers.

⑤⑧ 3 beams and a $14 saw came to $38. How much did each beam cost?

_____ y + _____ = _____

Each beam cost $_____ .

⑤⑨ 4 bags of sand and 20 kg of cement weigh 100 kg altogether, how heavy is one bag of sand?

_____ m + _____ = _____

One bag of sand is _____ kg.

⑥⓪ I work 8 hours a day and 5 days a week. If I earn $500 a week, how much do I earn an hour?

_____ x _____ p = _____

Uncle Fred earns $_____ an hour.

ACTIVITY

Circle the three correct number cards to solve Tony's riddle.

The sum of three numbers is 27. The second number is double the first. The third number is one less than the first. What are the numbers?

10	7	4
16	5	14
6	12	8

ISBN: 978-1-897164-21-1

11 Angles and Lines

Adjacent angles - the two angles that share a common side and vertex and do not overlap

Complementary angles - two adjacent angles that have a sum of 90°

Supplementary angles - two adjacent angles that have a sum of 180°

Opposite angles - the opposite and congruent angles formed by two intersecting lines

Transversal - a line intersecting two or more lines

Parallel lines - two or more lines that are in the same plane and do not intersect (symbol: ⤢⤢)

Transversal

Parallel lines

Alternate angles: ∠a = ∠d, ∠b = ∠c

Corresponding angles: ∠a = ∠e, ∠b = ∠f

Measure the angles and tell whether they are adjacent, complementary, supplementary or opposite.

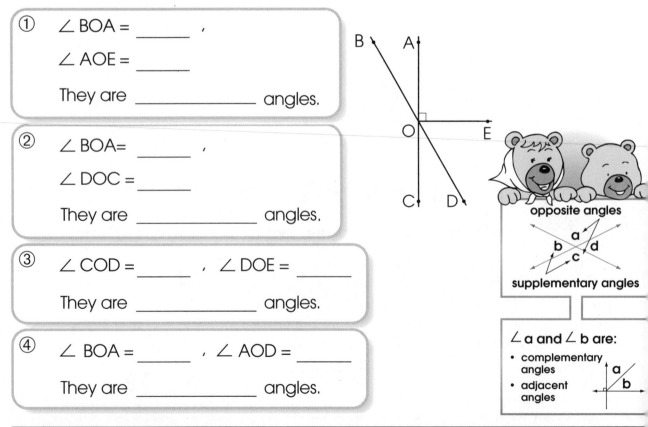

① ∠ BOA = _____ ,

∠ AOE = _____

They are _____ angles.

② ∠ BOA = _____ ,

∠ DOC = _____

They are _____ angles.

③ ∠ COD = _____ , ∠ DOE = _____

They are _____ angles.

④ ∠ BOA = _____ , ∠ AOD = _____

They are _____ angles.

opposite angles

supplementary angles

∠ a and ∠ b are:
- complementary angles
- adjacent angles

ISBN: 978-1-897164-21-1

Find the value of the unknown angles.

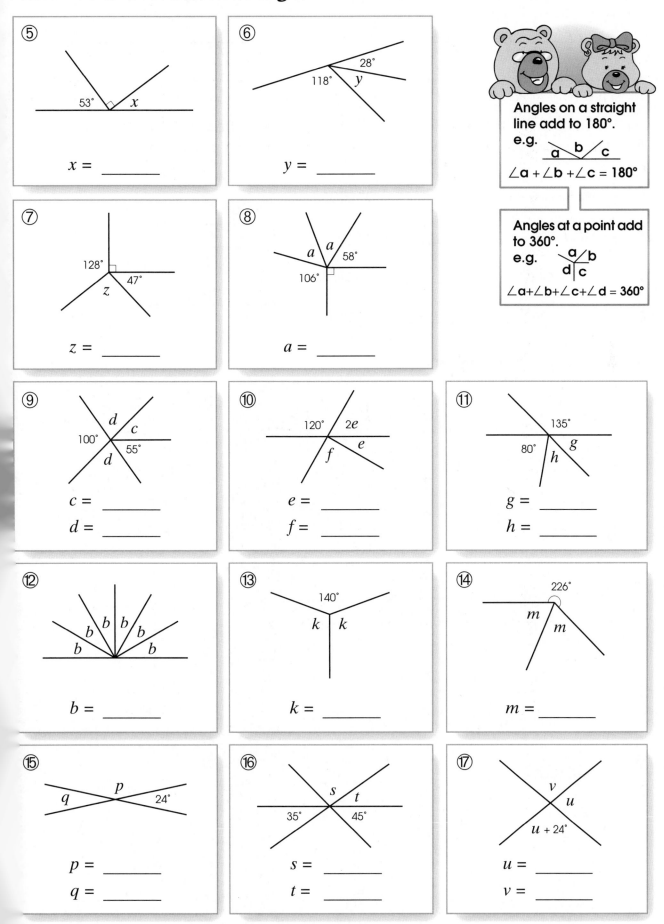

⑤

53° x

x = _____

⑥

28°
118° y

y = _____

ISBN: 978-1-897164-21-1

Angles on a straight line add to 180°.
e.g.
a b c
∠a + ∠b + ∠c = 180°

Angles at a point add to 360°.
e.g.
a b
d c
∠a + ∠b + ∠c + ∠d = 360°

⑦

128°
47°
z

z = _____

⑧

a a 58°
106°

a = _____

⑨

d c
100° 55°
d

c = _____

d = _____

⑩

120° 2e
f e

e = _____

f = _____

⑪

135°
80° g
h

g = _____

h = _____

⑫

b b b
b b

b = _____

⑬

140°
k k

k = _____

⑭

226°
m m

m = _____

⑮

p
q 24°

p = _____

q = _____

⑯

s t
35° 45°

s = _____

t = _____

⑰

v
u
u + 24°

u = _____

v = _____

Follow Tony's method to tell whether the angles are corresponding or alternate.

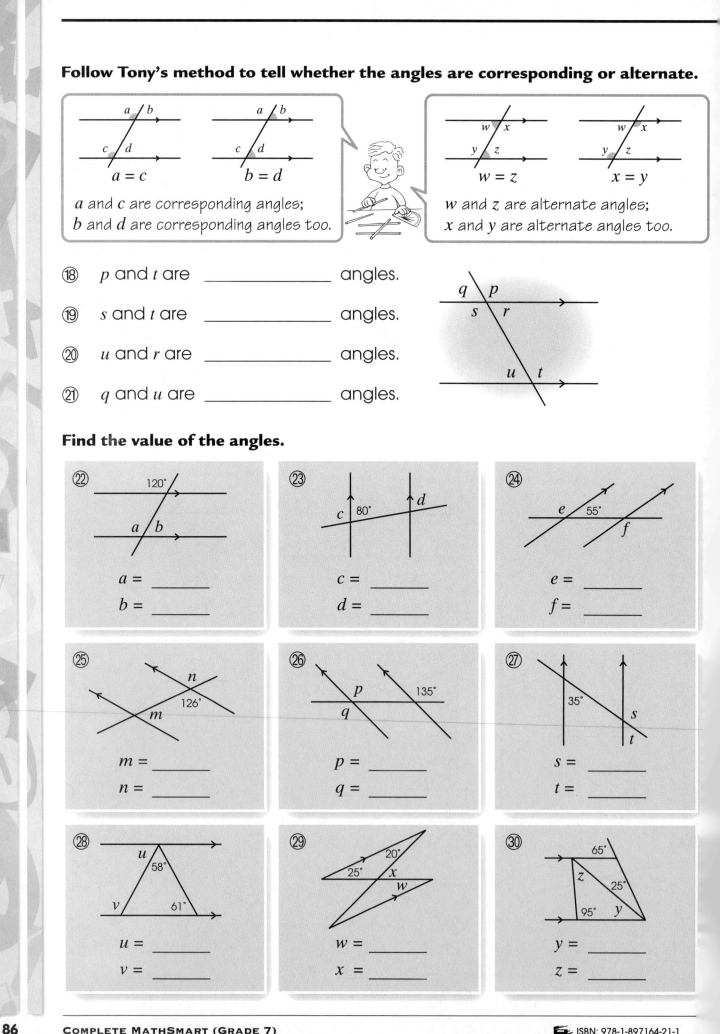

a and *c* are corresponding angles;
b and *d* are corresponding angles too.

w and *z* are alternate angles;
x and *y* are alternate angles too.

⑱ *p* and *t* are _____ angles.

⑲ *s* and *t* are _____ angles.

⑳ *u* and *r* are _____ angles.

㉑ *q* and *u* are _____ angles.

Find the value of the angles.

㉒ 120°
a = _____
b = _____

㉓ *c* 80°   *d*
c = _____
d = _____

㉔ *e* 55°   *f*
e = _____
f = _____

㉕ *n*   126°   *m*
m = _____
n = _____

㉖ *p* 135°   *q*
p = _____
q = _____

㉗ 35°   *s*   *t*
s = _____
t = _____

㉘ *u* 58°   *v* 61°
u = _____
v = _____

㉙ 20°   25°   *x*   *w*
w = _____
x = _____

㉚ 65°   *z*   25°   95°   *y*
y = _____
z = _____

ISBN: 978-1-897164-21-1

Follow the children's methods to find the marked angles.

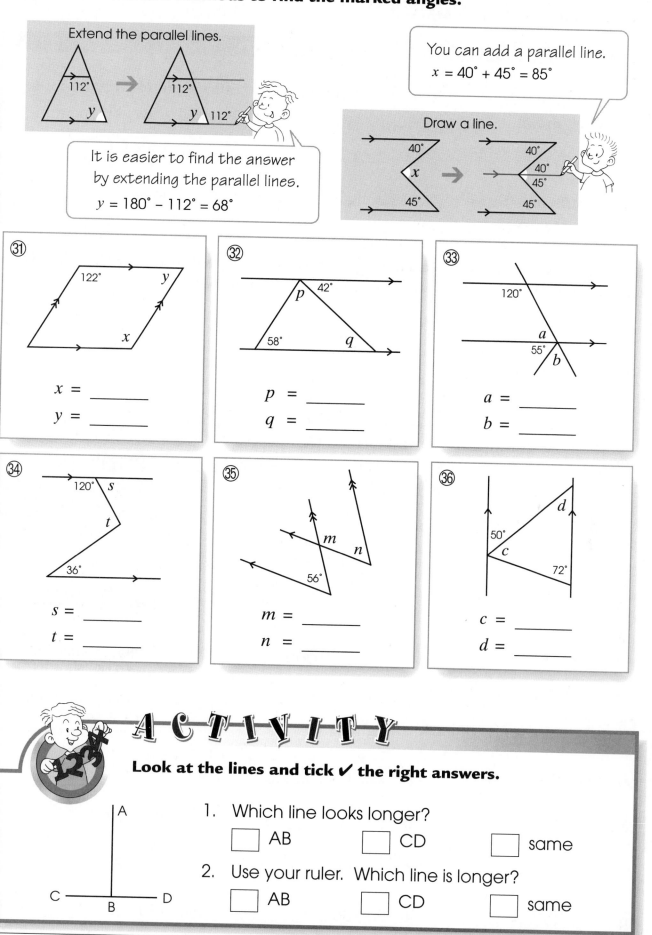

Extend the parallel lines.

112° → 112° y 112°

It is easier to find the answer by extending the parallel lines.

$y = 180° - 112° = 68°$

You can add a parallel line.

$x = 40° + 45° = 85°$

Draw a line.

40° x 45° → 40° 40° 45° 45°

---

③① 122° y x

$x = $ _____
$y = $ _____

③② 42° p 58° q

$p = $ _____
$q = $ _____

③③ 120° a 55° b

$a = $ _____
$b = $ _____

③④ 120° s t 36°

$s = $ _____
$t = $ _____

③⑤ m n 56°

$m = $ _____
$n = $ _____

③⑥ d 50° c 72°

$c = $ _____
$d = $ _____

ACTIVITY

Look at the lines and tick ✔ the right answers.

A

C ——— B ——— D

1. Which line looks longer?

☐ AB ☐ CD ☐ same

2. Use your ruler. Which line is longer?

☐ AB ☐ CD ☐ same

ISBN: 978-1-897164-21-1

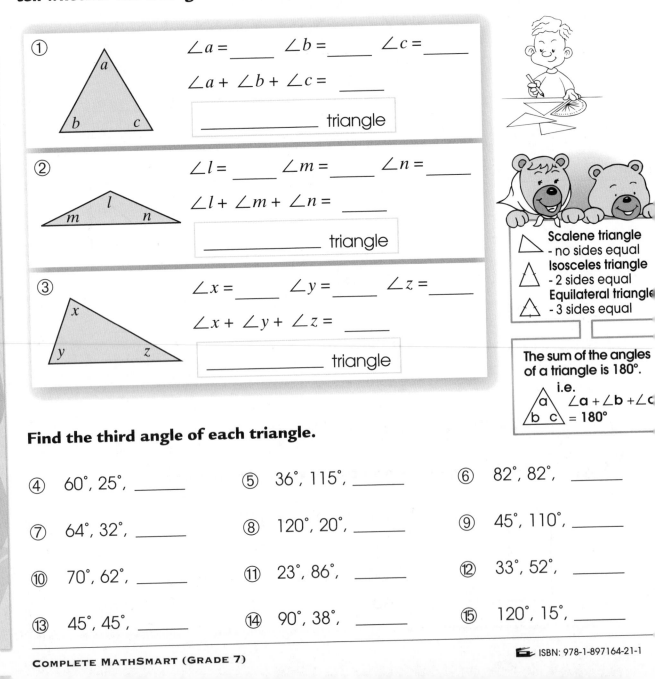

12 Angles and Shapes

Bisector — a line that divides an angle or a line into two equal parts

Line segment — a part of a line bounded and named by two end points

e.g.   A •————————• B   segment AB or $\overline{AB}$

Perpendicular lines — two lines that intersect to form right angles

Use a protractor to measure the angles in each triangle and find their sum. The
tell whether the triangle is scalene, isosceles or equilateral.

① $\angle a =$ _____ $\angle b =$ _____ $\angle c =$ _____

$\angle a + \angle b + \angle c =$ _____

_____ triangle

② $\angle l =$ _____ $\angle m =$ _____ $\angle n =$ _____

$\angle l + \angle m + \angle n =$ _____

_____ triangle

③ $\angle x =$ _____ $\angle y =$ _____ $\angle z =$ _____

$\angle x + \angle y + \angle z =$ _____

_____ triangle

Scalene triangle
- no sides equal
Isosceles triangle
- 2 sides equal
Equilateral triangle
- 3 sides equal

The sum of the angles
of a triangle is 180°.
i.e.
$\angle a + \angle b + \angle c = 180°$

Find the third angle of each triangle.

④ 60°, 25°, _____

⑤ 36°, 115°, _____

⑥ 82°, 82°, _____

⑦ 64°, 32°, _____

⑧ 120°, 20°, _____

⑨ 45°, 110°, _____

⑩ 70°, 62°, _____

⑪ 23°, 86°, _____

⑫ 33°, 52°, _____

⑬ 45°, 45°, _____

⑭ 90°, 38°, _____

⑮ 120°, 15°, _____

ISBN: 978-1-897164-21-1

Construct a triangle with sides 4 cm, 5 cm and 6 cm.

1st   Draw a line segment AB = 6 cm.
2nd   Take A as the centre and a compass radius of 5 cm; draw an arc.
3rd   Take B as the centre and a compass radius of 4 cm; draw
       an arc to cut the first arc at point C.
4th   Join AC and BC, and you can get △ABC.

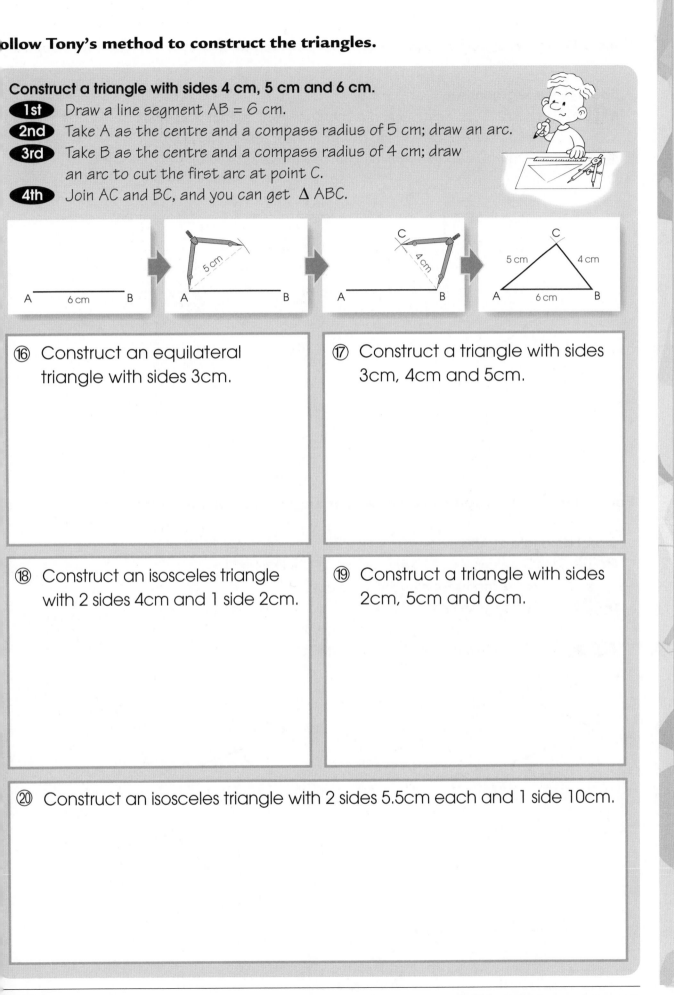

⑯ Construct an equilateral
   triangle with sides 3cm.

⑰ Construct a triangle with sides
   3cm, 4cm and 5cm.

⑱ Construct an isosceles triangle
   with 2 sides 4cm and 1 side 2cm.

⑲ Construct a triangle with sides
   2cm, 5cm and 6cm.

⑳ Construct an isosceles triangle with 2 sides 5.5cm each and 1 side 10cm.

ISBN: 978-1-897164-21-1

Follow Dave's method to construct a perpendicular bisector for each line segmen

Bisect a line segment $\overline{AB}$:

1st Use A and B as centres and a compass radius more than half the length of $\overline{AB}$, construct two arcs intersecting at C and D.

2nd Join C and D. $\overline{CD}$ is perpendicular to and bisects $\overline{AB}$.

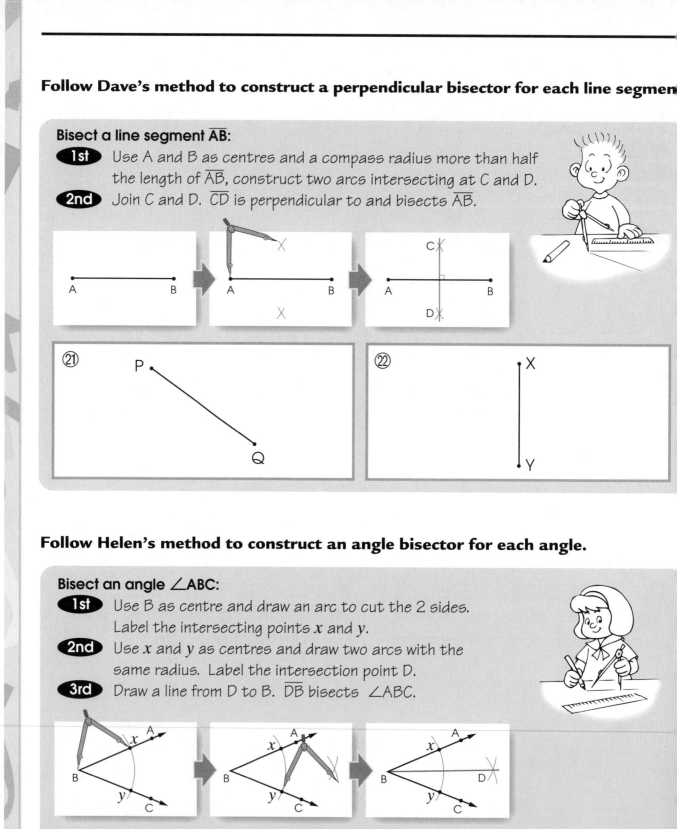

Follow Helen's method to construct an angle bisector for each angle.

Bisect an angle $\angle ABC$:

1st Use B as centre and draw an arc to cut the 2 sides. Label the intersecting points x and y.

2nd Use x and y as centres and draw two arcs with the same radius. Label the intersection point D.

3rd Draw a line from D to B. $\overline{DB}$ bisects $\angle ABC$.

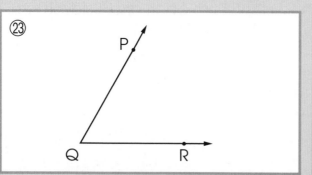

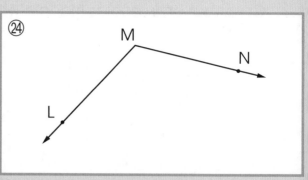

ISBN: 978-1-897164-21-1

Read what Tony says. Then tell whether each pair of triangles is congruent or similar.

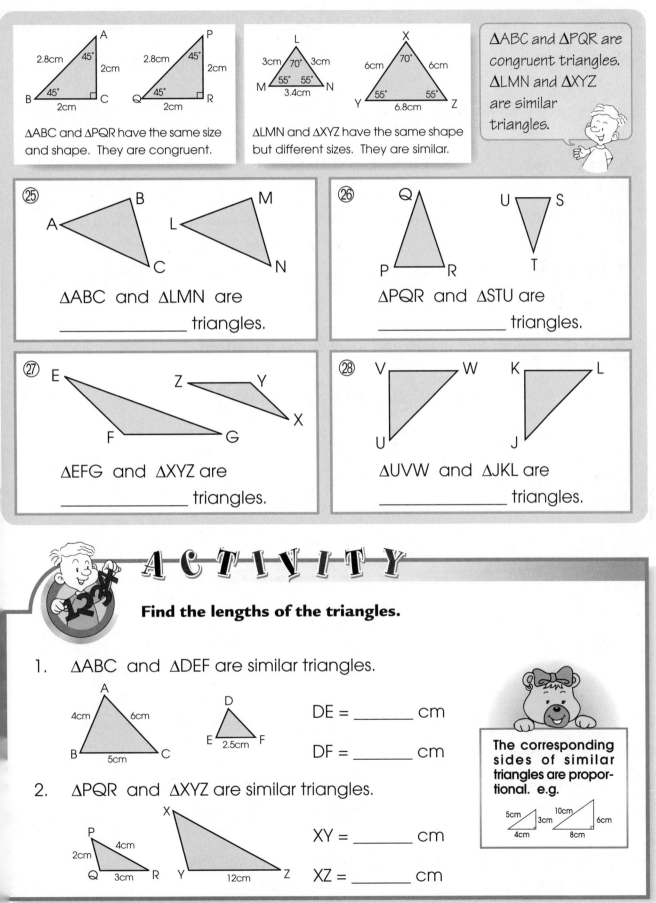

△ABC and △PQR have the same size and shape. They are congruent.

△LMN and △XYZ have the same shape but different sizes. They are similar.

△ABC and △PQR are congruent triangles. △LMN and △XYZ are similar triangles.

㉕ △ABC and △LMN are _____ triangles.

㉖ △PQR and △STU are _____ triangles.

㉗ △EFG and △XYZ are _____ triangles.

㉘ △UVW and △JKL are _____ triangles.

ACTIVITY

Find the lengths of the triangles.

1. △ABC and △DEF are similar triangles.

   DE = _____ cm

   DF = _____ cm

   The corresponding sides of similar triangles are proportional. e.g.

2. △PQR and △XYZ are similar triangles.

   XY = _____ cm

   XZ = _____ cm

ISBN: 978-1-897164-21-1

WORDS TO LEARN

Frequency	-	the number of times an event occurs within a period
Frequency distribution	-	a table showing the frequency of different groups of data
Circle graph	-	a graph using parts of a circle to show information about a whole
Histogram	-	a bar graph using connected bars to show the frequency of occurrence of grouped data
Broken-line graph	-	a graph made by joining successive plotted points

Look at the broken-line graph and answer the questions.

Weight of Collected Waste Paper

① How many more kilograms of waste paper were collected from Bridge Town than from Sunny Town in September? _____ kg

② How many more kilograms of waste paper were collected from Bridge Town in March than in August? _____ kg

③ In which month was the greatest increase in the weight of waste paper collected from Sunny Town? _____

④ In which month was there the greatest difference in weight between the waste paper collected from Bridge Town and that from Sunny Town? _____

Use the circle graph and your protractor to find the percent of waste paper collected from different towns.

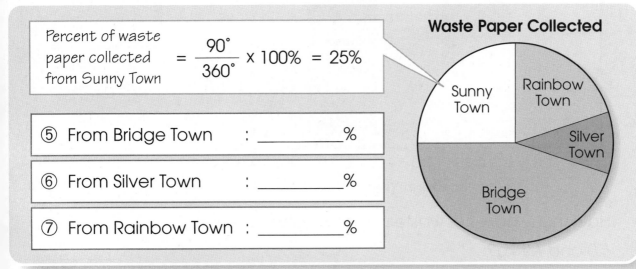

Percent of waste paper collected from Sunny Town $= \dfrac{90°}{360°} \times 100\% = 25\%$

Waste Paper Collected

⑤ From Bridge Town : _____ %

⑥ From Silver Town : _____ %

⑦ From Rainbow Town : _____ %

Follow Dave's method to find the size of angle for each item and complete the circle graph.

Paper for recycling $= 30\%$ of $360°$
$= 0.3 \times 360°$
$= 108°$

⑫ **Items for Recycling**

Paper

Items for Recycling	Percent	Angle
Paper	30 %	108°
⑧ Glass	20 %	
⑨ Plastic	25 %	
⑩ Cans	15 %	
⑪ Outside waste	10 %	

ISBN: 978-1-897164-21-1

Use the histogram to answer the questions.

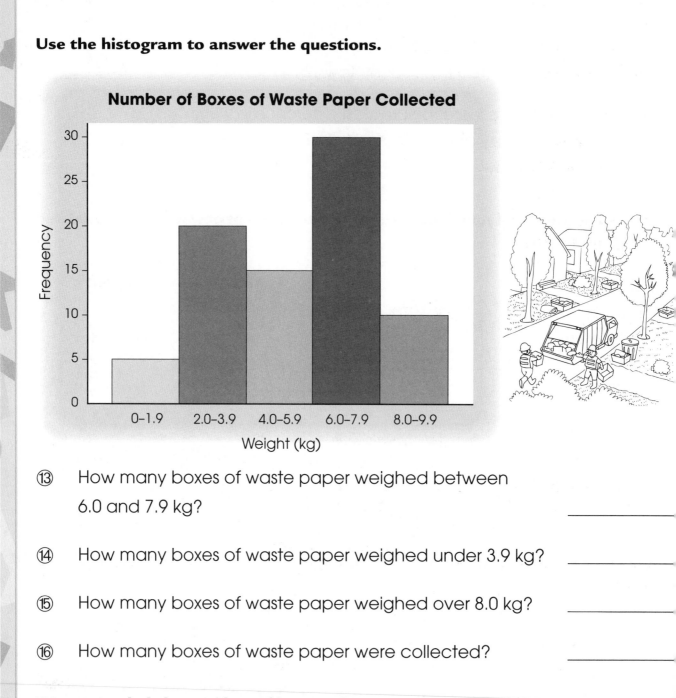

Number of Boxes of Waste Paper Collected

⑬ How many boxes of waste paper weighed between 6.0 and 7.9 kg? _____

⑭ How many boxes of waste paper weighed under 3.9 kg? _____

⑮ How many boxes of waste paper weighed over 8.0 kg? _____

⑯ How many boxes of waste paper were collected? _____

Tony recorded the number of boxes of waste paper collected weekly in his area. Use the data to complete the frequency distribution table and the histogram. Then answer the questions.

Number of boxes of waste paper collected weekly

68	77	61	80	71	70	78	56	78	75
62	67	73	77	80	79	69	78	75	66
70	80	59	80	72	65	77	68	79	79

ISBN: 978-1-897164-21-1

⑰

Number of boxes	56-60	61-65	66-70	71-75	76-80
Frequency	2				

⑱

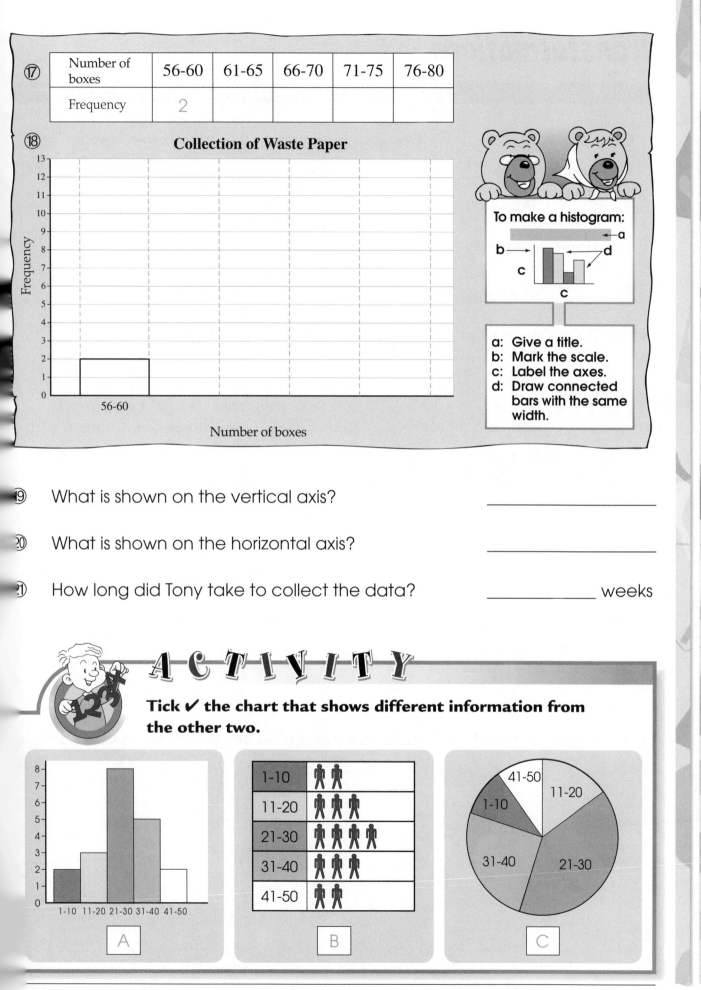

Collection of Waste Paper

To make a histogram:

b →
c
← a
← d
c

a: Give a title.
b: Mark the scale.
c: Label the axes.
d: Draw connected bars with the same width.

⑲ What is shown on the vertical axis? _____

⑳ What is shown on the horizontal axis? _____

㉑ How long did Tony take to collect the data? _____ weeks

ACTIVITY

Tick ✔ the chart that shows different information from the other two.

A

B

1-10	👤 👤
11-20	👤 👤 👤
21-30	👤 👤 👤 👤
31-40	👤 👤 👤
41-50	👤 👤

C

ISBN: 978-1-897164-21-1

Transformations

Translation	-	sliding each point of a plane in the same direction and distance
Rotation	-	turning the points of a plane about a fixed point
Reflection	-	flipping the points of a plane over a line
Rotational symmetry	-	a figure has rotational symmetry if its image fits onto itself in less than one full turn

Follow Dave's method to draw each reflection image in line *l*.

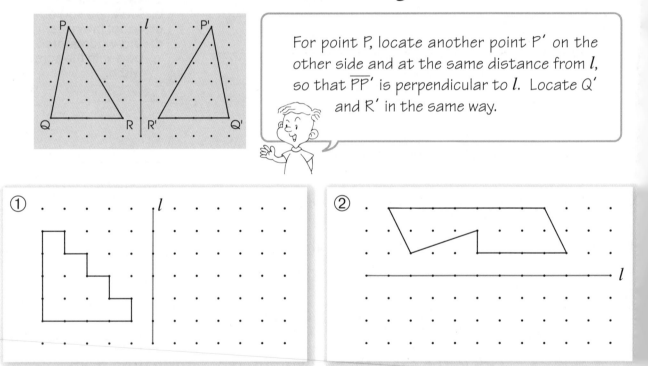

For point P, locate another point P' on the other side and at the same distance from *l*, so that $\overline{PP'}$ is perpendicular to *l*. Locate Q' and R' in the same way.

Follow Elaine's method to describe the translation images of the shaded triangle

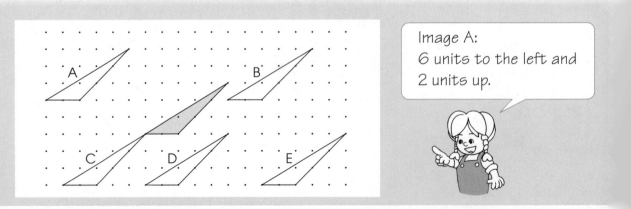

Image A:
6 units to the left and 2 units up.

   ISBN: 978-1-897164-21-1

③ Image B : _____

④ Image C : _____

⑤ Image D : _____

⑥ Image E : _____

Each triangle below is a rotation image of the shaded figure. Find the coordinates of the turning point, angle and direction of rotation.

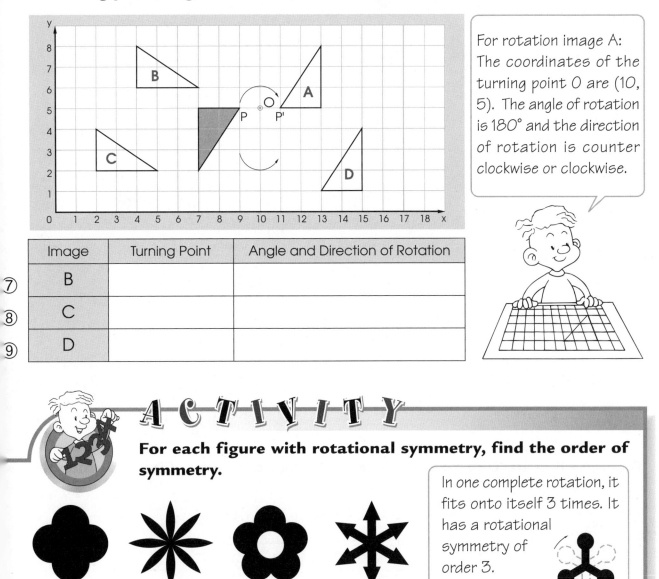

For rotation image A:
The coordinates of the turning point O are (10, 5). The angle of rotation is 180° and the direction of rotation is counter clockwise or clockwise.

	Image	Turning Point	Angle and Direction of Rotation
⑦	B		
⑧	C		
⑨	D		

ACTIVITY

For each figure with rotational symmetry, find the order of symmetry.

In one complete rotation, it fits onto itself 3 times. It has a rotational symmetry of order 3.

A   B   C   D

Shape	A	B	C	D
Order of Symmetry				

ISBN: 978-1-897164-21-1

15 Probability

WORDS TO LEARN

Mutually exclusive events - events which do not have common outcomes

Independent events - events with no effect on one another

Read what Uncle Fred says. Then help Tony find the probabilities.

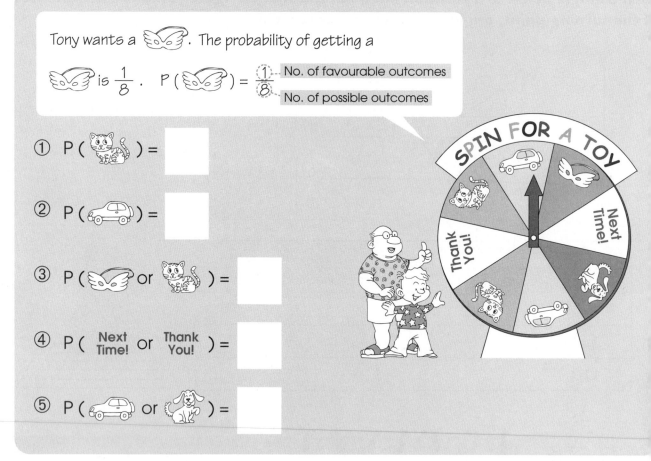

Tony wants a 🪽. The probability of getting a

🪽 is $\frac{1}{8}$. P (🪽) = $\frac{1}{8}$ — No. of favourable outcomes

— No. of possible outcomes

① P (🐱) =

② P (🚗) =

③ P (🪽 or 🐱) =

④ P (Next Time! or Thank You!) =

⑤ P (🚗 or 🐕) =

There are 8 cards. One card is drawn at random each time. Find the probabilities

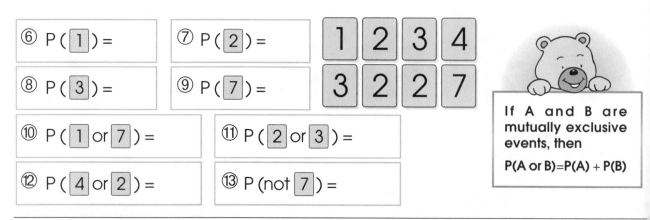

⑥ P (1) =   ⑦ P (2) =

⑧ P (3) =   ⑨ P (7) =

⑩ P (1 or 7) =   ⑪ P (2 or 3) =

⑫ P (4 or 2) =   ⑬ P (not 7) =

1 2 3 4
3 2 2 7

If A and B are mutually exclusive events, then

P(A or B)=P(A) + P(B)

ISBN: 978-1-897164-21-1

Follow Elaine's method to make a tree diagram and find the number of possible outcomes. Then find the probabilities.

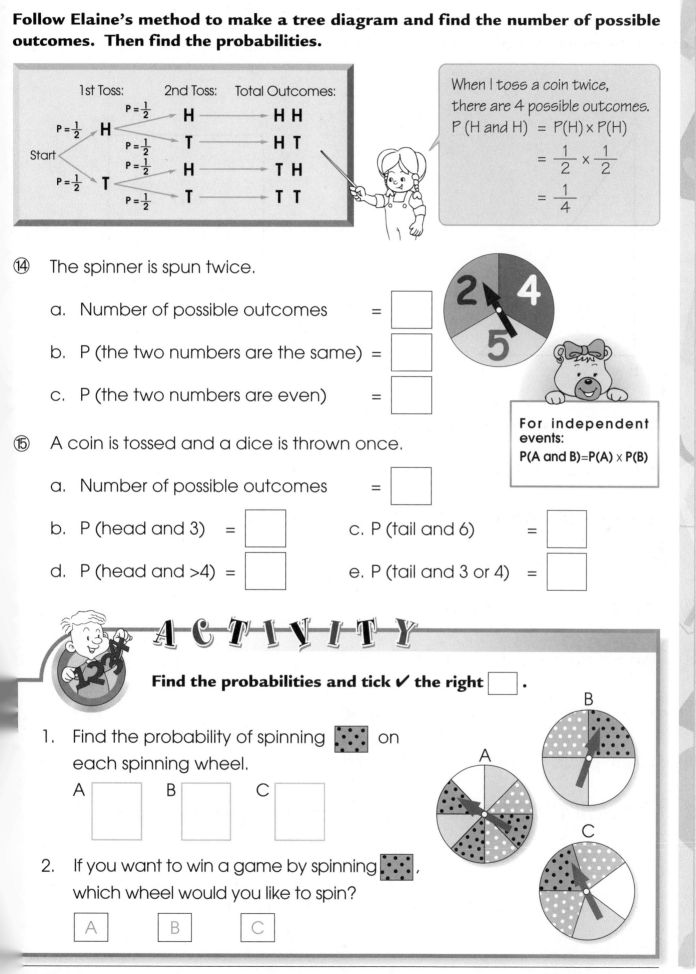

1st Toss: 2nd Toss: Total Outcomes:

Start

$P=\frac{1}{2}$ H
$P=\frac{1}{2}$ T

$P=\frac{1}{2}$ H → H H
$P=\frac{1}{2}$ T → H T
$P=\frac{1}{2}$ H → T H
$P=\frac{1}{2}$ T → T T

When I toss a coin twice, there are 4 possible outcomes.

$P(H \text{ and } H) = P(H) \times P(H)$

$= \frac{1}{2} \times \frac{1}{2}$

$= \frac{1}{4}$

⑭ The spinner is spun twice.

a. Number of possible outcomes = ☐

b. P (the two numbers are the same) = ☐

c. P (the two numbers are even) = ☐

For independent events:

$P(A \text{ and } B) = P(A) \times P(B)$

⑮ A coin is tossed and a dice is thrown once.

a. Number of possible outcomes = ☐

b. P (head and 3) = ☐   c. P (tail and 6) = ☐

d. P (head and >4) = ☐   e. P (tail and 3 or 4) = ☐

ACTIVITY

Find the probabilities and tick ✔ the right ☐.

1. Find the probability of spinning ▦ on each spinning wheel.

   A ☐   B ☐   C ☐

2. If you want to win a game by spinning ▦, which wheel would you like to spin?

   ☐ A   ☐ B   ☐ C

ISBN: 978-1-897164-21-1

Look at the coordinate plane and answer the questions. (15 marks)

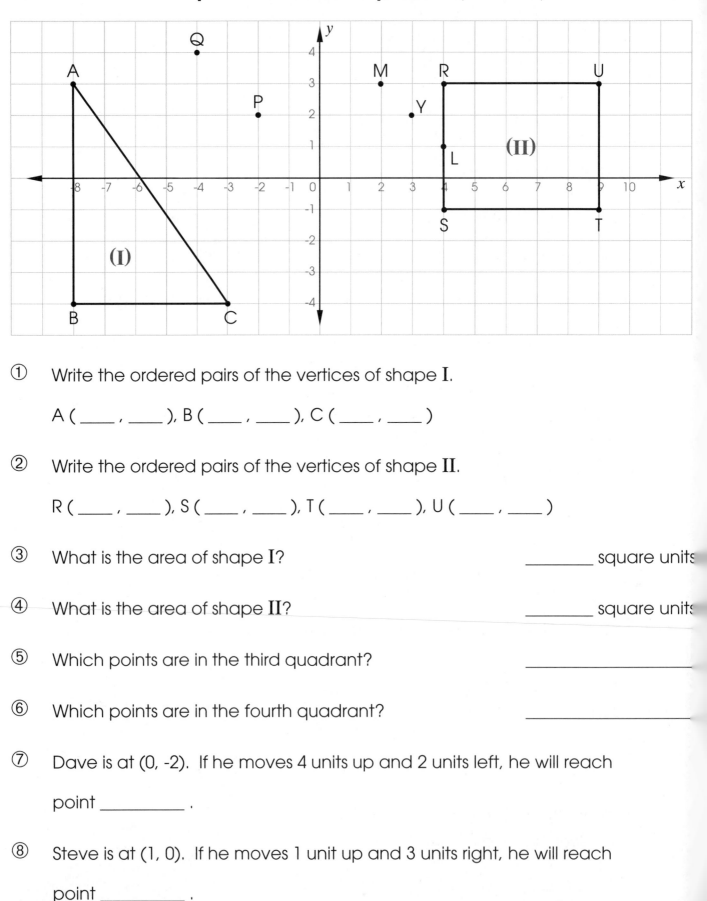

① Write the ordered pairs of the vertices of shape I.

A (____ , ____), B (____ , ____), C (____ , ____)

② Write the ordered pairs of the vertices of shape II.

R (____ , ____), S (____ , ____), T (____ , ____), U (____ , ____)

③ What is the area of shape I?   _____ square units

④ What is the area of shape II?   _____ square units

⑤ Which points are in the third quadrant?   _____

⑥ Which points are in the fourth quadrant?   _____

⑦ Dave is at (0, -2). If he moves 4 units up and 2 units left, he will reach

point _____ .

⑧ Steve is at (1, 0). If he moves 1 unit up and 3 units right, he will reach

point _____ .

ISBN: 978-1-897164-21-1

Write in algebraic expressions. (4 marks)

⑨ The sum of 4 and the square of x.

_____

⑩ Subtract 16 from the product of y and 4.

_____

⑪ 3 more than m times 9.

_____

⑫ The difference of q and 5, divided by 4.

_____

Evaluate for p = 6 and q = -3. (4 marks)

⑬ $4p + 2q$ = _____

⑭ $-2pq$ = _____

⑮ $(3p - 4) \times q$ = _____

⑯ $p^2 + q^2$ = _____

Simplify the expressions. (4 marks)

⑰ $3(a - b) + 4b$

= _____

⑱ $2(8 - m) + 5m$

= _____

⑲ $5(n + 7) - 2(n + 1)$

= _____

⑳ $2(t + 6) - 3(2t - 1)$

= _____

Solve the equations. (4 marks)

㉑ $6a - 3a - 0.5a = 9$

㉒ $3.5b - 20 = 11.5$

ISBN: 978-1-897164-21-1

Find the value of the angles. (16 marks)

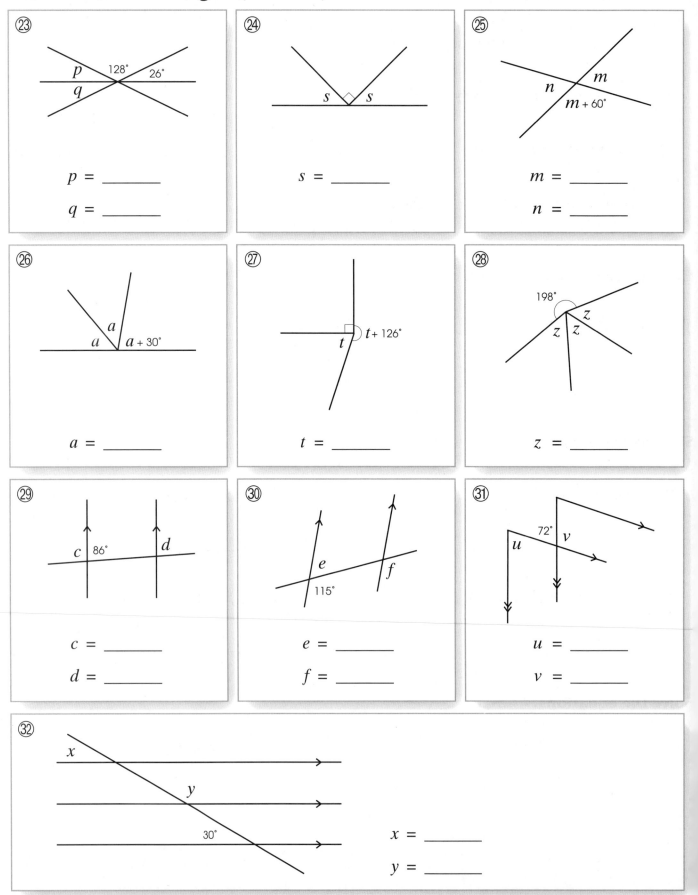

㉓
p 128° 26°
q

$p =$ _____

$q =$ _____

㉔
s s

$s =$ _____

㉕
n m
$m + 60°$

$m =$ _____

$n =$ _____

㉖
a
a $a + 30°$

$a =$ _____

㉗
t $t + 126°$

$t =$ _____

㉘
198°
z
z z

$z =$ _____

㉙
c 86° d

$c =$ _____

$d =$ _____

㉚
e
115° f

$e =$ _____

$f =$ _____

㉛
72° v
u

$u =$ _____

$v =$ _____

㉜
x
y
30°

$x =$ _____

$y =$ _____

Write congruent or similar. (3 marks)

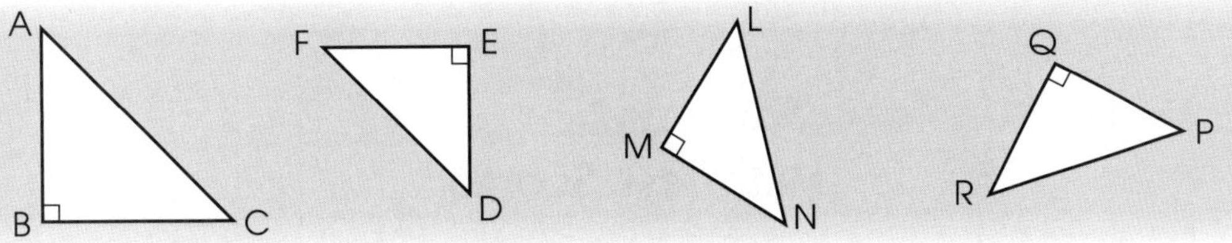

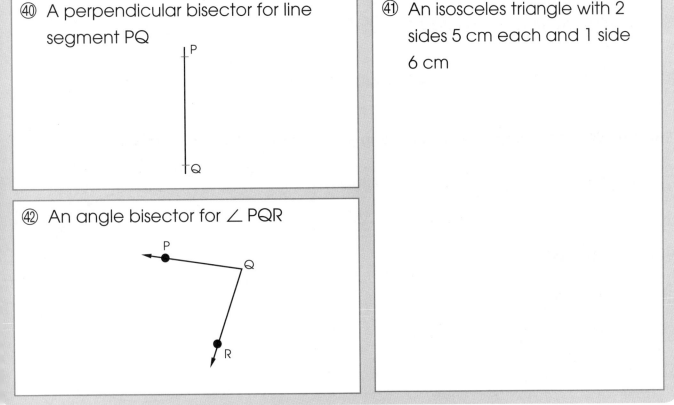

㉝ △ ABC and △ DEF are ________________ triangles.

㉞ △ LMN and △ PQR are ________________ triangles.

㉟ △ DEF and △ PQR are ________________ triangles.

Find the third angle of each triangle. (4 marks)

㊱ 33°, 46°, ________

㊲ 120°, 36°, ________

㊳ 98°, 37°, ________

㊴ 49°, 52°, ________

Construct the following with a ruler and a pair of compasses only. (6 marks)

㊵ A perpendicular bisector for line segment PQ

㊶ An isosceles triangle with 2 sides 5 cm each and 1 side 6 cm

㊷ An angle bisector for ∠ PQR

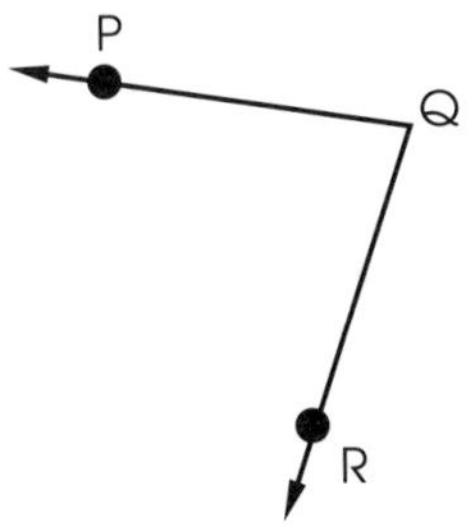

Use the histogram to complete the frequency distribution table and answer the questions.
(8 marks)

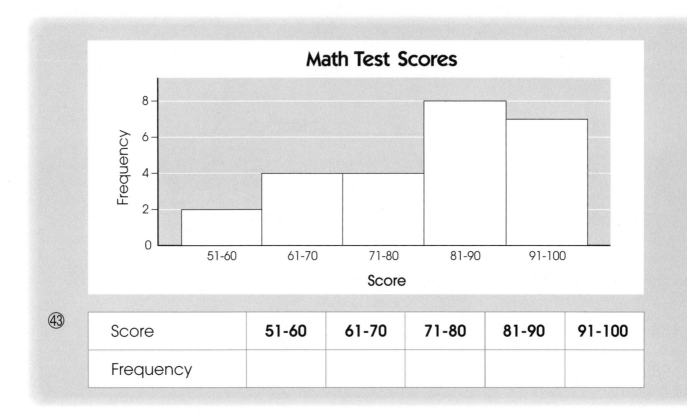

Math Test Scores

④③

Score	51-60	61-70	71-80	81-90	91-100
Frequency					

④④ How many students scored between 61 and 80? _____

④⑤ How many students scored over 81? _____

④⑥ How many students participated in the test? _____

Find the size of the angle for each group and complete the circle graph. (8 marks)

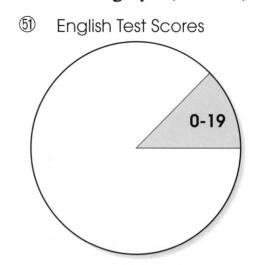

English Test Scores	Percent	Angle
0 - 19	12.5 %	45°
④⑦ 20 - 39	12.5 %	
④⑧ 40 - 59	30 %	
④⑨ 60 - 79	25 %	
⑤⓪ 80 - 100	20 %	

⑤① English Test Scores

0-19

ISBN: 978-1-897164-21-1

Draw each reflection image in the line *l*. (4 marks)

(52)

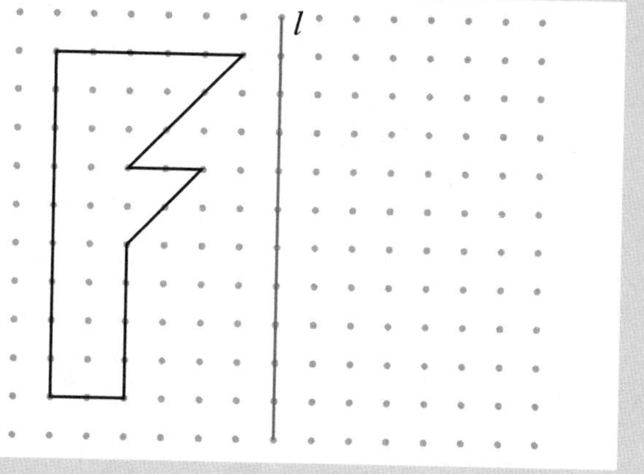

(53)

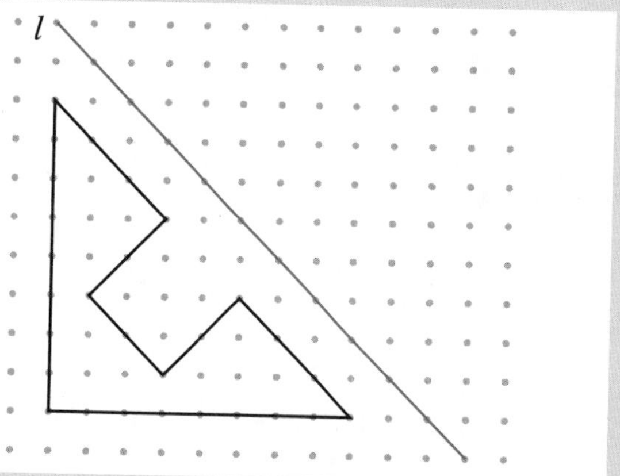

Describe the translation images of the shaded trapezoid. Write the numbers and tick ✔ the right boxes. (4 marks)

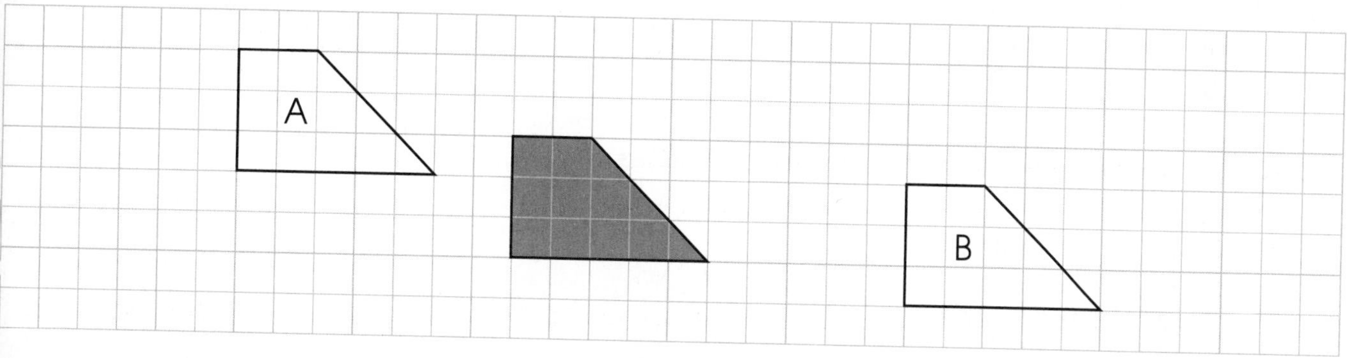

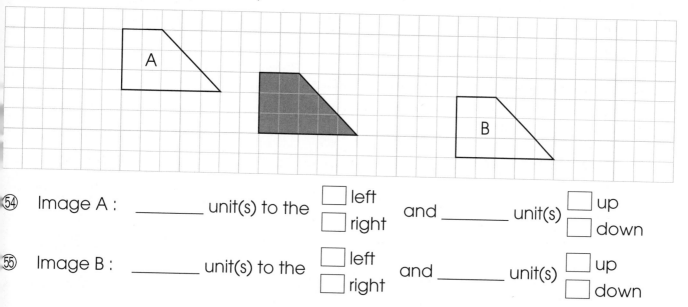

(54) Image A : ________ unit(s) to the ☐ left ☐ right and ________ unit(s) ☐ up ☐ down

(55) Image B : ________ unit(s) to the ☐ left ☐ right and ________ unit(s) ☐ up ☐ down

Find the coordinates of the turning point, angle and direction of rotation of each rotation image of the shaded triangle. (4 marks)

	Image	Turning Point	Angle and Direction of Rotation
(56)	A	(,)	
(57)	B	(,)	

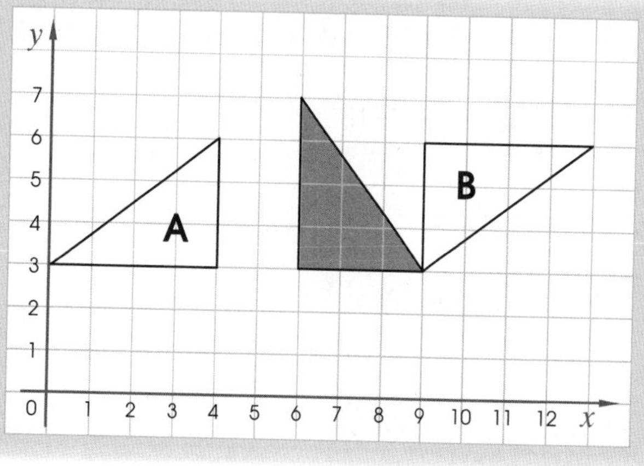

ISBN: 978-1-897164-21-1

Use the spinner below to find the probabilities. (6 marks)

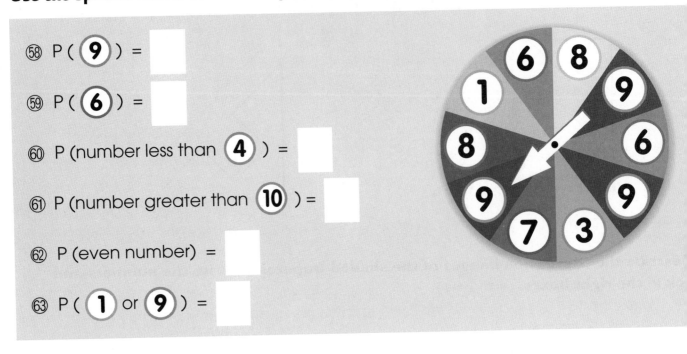

⑤⑧ P (⑨) = ☐

⑤⑨ P (⑥) = ☐

⑥⓪ P (number less than ④) = ☐

⑥① P (number greater than ⑩) = ☐

⑥② P (even number) = ☐

⑥③ P (① or ⑨) = ☐

List all the possible outcomes and probabilities. (6 marks)

⑥④ The spinner is spun twice.

   a. Possible outcomes : _____

   _____

   _____

   b. P (two letters are the same) = _____

   c. P (one letter is **B**) = _____

⑥⑤ A coin is tossed 3 times. (H for head and T for tail)

   a. Possible outcomes : _____

   _____

   b. P (exactly 2 heads) = _____

   c. P (at least 2 tails) = _____

SCORE

10

ISBN: 978-1-897164-21-1

Section III

Overview

Section II allows students to further practise and develop important mathematical concepts. In this section, they will use these concepts to solve word problems as well as problems that they will encounter in real-life situations.

In Number Sense and Numeration, students need to apply all four arithmetic operations to fractions, decimals, integers, exponents, squares, and square roots to solve problems. The order of operation concept is reinforced. The use of percents in calculating simple interest, discounts, and tax is also included.

In Patterning and Algebra, students will examine patterns and trends to predict outcomes. They will also practise writing and solving equations to represent and solve problems.

Order of Operations

Sequence of Operations

$4 - 2 + 5 \times 4^2 \div (5 + 3) \rightarrow$ 1st **B**rackets

$= 4 - 2 + 5 \times 4^2 \div 8 \rightarrow$ 2nd **E**xponents

$= 4 - 2 + 5 \times 16 \div 8 \rightarrow$ 3rd **D**ivide or

$= 4 - 2 + 80 \div 8$ **M**ultiply (whichever comes first)

$= 4 - 2 + 10 \rightarrow$ 4th **A**dd or

$= 12$ **S**ubtract (whichever comes first)

Solve the problems. Show your work.

① Debbie bought 3 pairs of earrings on sale. The normal price is $6.00 a pair but they are on sale for $1.00 off. Write an expression to find the final cost of her purchase in dollars.

Answer : The final cost of her purchase is _____ dollars. _____

② Joy buys 3 CDs at $19.00 each (including tax). She has a $5.00 off coupon and her father agrees to pay half the cost. Write an expression to show the total amount Joy pays.

Answer : _____

③ Pat has an old calculator. She punches in $10 + 12 \div 2$. The calculator calculates following the sequence of the punches. What answer does it give? Is it right or wrong?

Answer : _____

④ What error has Pat's calculator made?

Answer : _____

⑤ What is the correct answer for the expression that Pat punched in?

Answer : _____

     ISBN: 978-1-897164-21-1

⑥ Jim says, 'The sum of the squares of two numbers is the same as the square of the sum of the numbers.' Do you agree with him? Write Yes or No. Then illustrate your answer with an example.

Answer : _____

⑦ Judy has a weekend job which pays $7.00 per hour and $10.00 per hour overtime. One week before Christmas, Judy worked 6 hours on Saturday and 5 hours on Sunday plus 4 hours overtime. In addition she received a bonus of $20.00. How much did she earn?

Answer : _____

⑧ Write two different expressions for the perimeter of the rectangle, one with brackets and the other without. Show that the answers are the same.

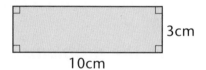

10cm   3cm

Answer : _____

⑨ Andrew has written a mathematical statement : 35 + 3 ÷ 19 + 7 × 22 − 13 = 65. Rewrite the statement and add brackets to make it correct.

Answer : _____

CHALLENGE

Mrs Wing issues a challenge to her Grade 7 Math class. 'Use all the digits 1-9 inclusive once only and any mathematical operations to produce the answer 200.' Janet gives an answer 200 = 2 × (1 + 9) × (3 + 7) × (8 − 5 + 4 − 6). Try to find another calculation to produce the answer 200.

Answer : _____

EXAMPLE

The sum (+) of two numbers is 20. The difference (−) between the numbers is 6. Determine the product (×) of the two numbers.

Solve the problem by systematic trial:

The bigger number must be more than half of 20, i.e. 10. So try 11.

$11 + 9 = 20$	but	$11 − 9 \neq 6$	(incorrect)
$12 + 8 = 20$	but	$12 − 8 \neq 6$	(incorrect)
$13 + 7 = 20$	and	$13 − 7 = 6$	(correct)

Answer : The numbers are 13 and 7, and the product is $13 \times 7 = 91$.

Solve the problems by systematic trial.

① Jane has made 2 long distance calls. One lasted 10 minutes longer than the other. The 2 calls lasted a total of 46 minutes. How long was each call?

Answer : _____

② Bill buys 2 pairs of jeans. He pays $15.00 more for one pair than the other. He pays with a $100 bill and gets $25.00 change. What is the price of the more expensive pair?

Answer : _____

③ The sum of 2 consecutive numbers is 933. What are the two numbers?

Answer : _____

④ The difference between the squares of 2 consecutive numbers is 25. Find the numbers.

Answer : _____

⑤ Find all 2-digit numbers which have a remainder of 2 when divided by 3, and a remainder of 3 when divided by 5.

Answer : _____

ISBN: 978-1-897164-21-1

⑥ Ann has 100 m of fencing. What is the largest possible straight-sided area that she can enclose on all 4 sides with the fencing?

Answer :

⑦ 20 students planned to go on a field trip splitting the cost of the bus among them. Then 5 more students decided to join in. As a result each student paid $5.00 less for the bus. Determine the total cost of the bus.

Answer :

Solve the problems. Show your work.

⑧ Tom buys 3 identical T-shirts in a sale. Each has been marked down by $5.00. He also buys a pair of jeans for $55.00. The total cost before tax is $106.00. Determine the price of each T-shirt before the sale.

Answer :

⑨ A pilot flies 1200 m below the clouds. Another plane at a lower altitude flies 1800 m above the ground. The clouds are 4000 m above the ground. What is the difference in altitude between the 2 planes?

Answer :

⑩ At the Richmond Nursery, trees are grown in square plots that are side by side. The side of each square is a 1 m section of fence.

a.   What is the total length of fence used to form a row of 30 tree plots?

Answer :

b.   What is the total length of fence used to form a row of 40 tree plots?

Answer :

ISBN: 978-1-897164-21-1

⑪ Ron and Fred run a 42 km marathon. When Ron has run twice as far as the distance he has already covered, he will be 2 km from the end.

a. How far has he run already?

Distance = Speed × Time ← Read this first.

$$Speed = \frac{Distance}{Time}$$

$$Time = \frac{Distance}{Speed}$$

Answer : He has run _____ .

b. If Ron has been running for 2 hours, what is his average speed so far?

Answer : _____

c. If Fred runs at a speed of 7 km per hour, how long will he take to complete the race?

Answer : _____

d. After 3 hours of running, how far is Ron ahead of Fred?

Answer : _____

e. Ron slows down after running for 3 hours. If he can complete the race in 5 hours, what is his average speed in the last 2 hours?

Answer : _____

f. Fred gives up after he has run for 5 hours. How far is he from the end?

Answer : _____

ISBN: 978-1-897164-21-1

⑫ A loonie has a diameter of approximately 27 mm and a thickness of about 2 mm.

    a.   How many loonies would it take, placed edge to edge, to cover 1 km?

Answer : _____

    b.   How many loonies would there be in a pile 1 m high?

Answer : _____

⑬ It is 12:35 p.m. now. What time was it 100 minutes ago?

Answer : _____

⑭ How many minutes will there be in the year 2002?

Answer : _____

⑮ Would the number of minutes in the year 2002 and 2004 be the same? Explain.

Answer : _____

⑯ At Hill School, there are 200 Grade 7 students and 160 of them have brown eyes. 90 have both brown hair and brown eyes.

    a.   How many Grade 7 students do not have brown eyes?

Answer : _____

    b.   How many brown-eyed Grade 7 students do not have brown hair?

Answer : _____

⑰ Mandy's heart rate is 70 beats per minute. If she lives to be exactly 90 years old, how many times will her heart beat throughout her life? (Assume there are no leap years.)

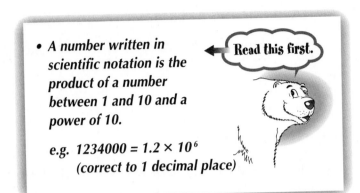

- A number written in scientific notation is the product of a number between 1 and 10 and a power of 10.

e.g. $1234000 = 1.2 \times 10^6$ (correct to 1 decimal place)

Answer : _____

⑱ Calculate the number of heartbeats throughout Mandy's life and show the answer in scientific notation. (correct to 1 decimal place)

Answer : _____

⑲ In 1997, Microsoft was worth 190 billion dollars. Mike showed this value with 3 scientific notations: a) 19×10^{10} b) 1.9×10^{11} and c) 0.19×10^{12}. Which one is correct? Explain.

Answer : _____

⑳ The average of 6 numbers is 5×10^6. The average of the first 5 numbers is 6×10^6. What is the last number?

Answer : _____

㉑ How many kilometres are there in 4.6×10^{10} cm?

Answer : _____

㉒ An aquarium is half-filled with water. Every minute, 100 mL of water is pumped through the filter into the aquarium. It takes 2 hours to fill up the aquarium. What is the capacity of the aquarium?

Answer : _____

ISBN: 978-1-897164-21-1

㉓ There are 32 teams in a hockey tournament. A team is eliminated if it loses a game.

a. How many games must be played in the tournament until the winner is determined?

Answer : _____

b. What if there are 64 teams?

Answer : _____

c. What if there are 128 teams? Describe the pattern in these answers.

Answer : _____

CHALLENGE

① Determine, without using a calculator, the value of
$(2 + 4 + 6 ... + 4800) - (1 + 3 + 5 ... + 4799)$

Answer : _____

② Jane drove from Antville to Beechwood and back, a distance of 300 km each way. The total journey there and back took 9 hours. She drove back 15 km/h faster than she drove there. Determine the average speed in each direction.

Answer : From Antville to Beechwood : _____

From Beechwood to Antville : _____

ISBN: 978-1-897164-21-1

UNIT 3

Squares and Square Roots

EXAMPLE

a. Find the value of $\sqrt{225}$.

The square root of N ($\sqrt{N}$) means the number which, when multiplied by itself, gives N.

so $\sqrt{225} = \sqrt{15 \times 15} = 15$

Answer : The value of $\sqrt{225}$ is 15.

b. Find the value of 13^2.

The square of N (N^2) means that N has to be multiplied by itself.

so $13^2 = 13 \times 13 = 169$.

Answer : The value of 13^2 is 169.

Solve the problems. Show your work.

① Estimate and put the following numbers on the number line below (without using a calculator):

$\sqrt{15}$, 3^2, $\sqrt{10}$, 1^2, $(\frac{1}{2})^2$, $\sqrt{\frac{1}{4}}$, $\sqrt{50}$

 Answer :

0   1   2   3   4   5   6   7   8   9   10

② Explain the difference between $\sqrt{9}$ and 9^2.

Answer : _____

③ State True or False for each of the following:

a. $\sqrt{3} \times \sqrt{5} = \sqrt{15}$

b. $\sqrt{4 + 9} = \sqrt{4} + \sqrt{9}$

c. $\sqrt{(81 + 9)^2} = 81 + 9$

d. $\sqrt{4^2 + 3^2} = 4 + 3$

Answer : a. _____ b. _____ c. _____ d. _____

ISBN: 978-1-897164-21-1

④ List the first 10 perfect squares. Describe the pattern in the difference between the numbers.

- A perfect square is a number with a whole number square root, e.g. 144 is a perfect square because

$$\sqrt{144} = \sqrt{12 \times 12} = 12$$

Read this first.

Answer : _____

⑤ If you multiply a perfect square by a perfect square, is the answer always a perfect square? Give an example to illustrate your answer.

Answer : _____

⑥ Brad pours a foundation for a square garden shed. The area of the foundation is 9 m^2. How many metres of lumber does Brad need to frame the whole foundation?

Answer : _____

⑦ The Great Pyramid at Khufu in Egypt has a square base that covers about $53\,000 \text{ m}^2$. About how long is each side of the base?   (correct to the nearest m)

Answer : _____

⑧ What is the difference in length of the sides of two squares if their areas are 121 m^2 and 289 m^2 ?

Answer : _____

CHALLENGE

2   3

3

2

Determine the shaded area and draw a square with an area equal to the shaded area.

Answer : _____

ISBN: 978-1-897164-21-1

UNIT 4 Exponents

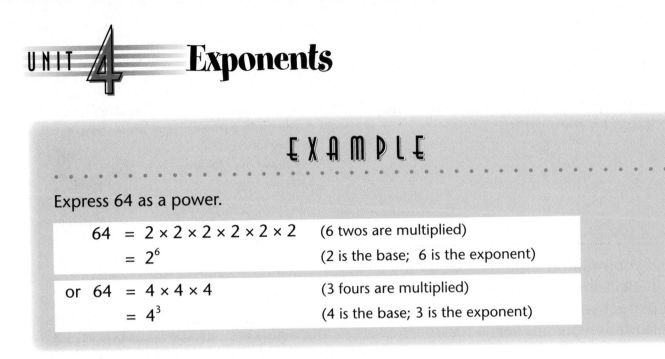

EXAMPLE

Express 64 as a power.

$$64 = 2 \times 2 \times 2 \times 2 \times 2 \times 2 \quad \text{(6 twos are multiplied)}$$
$$= 2^6 \quad \text{(2 is the base; 6 is the exponent)}$$

$$\text{or } 64 = 4 \times 4 \times 4 \quad \text{(3 fours are multiplied)}$$
$$= 4^3 \quad \text{(4 is the base; 3 is the exponent)}$$

Solve the problems. Show your work.

① Explain the difference between 5^4 and 5×4.

Answer : _____

② Which is bigger, 3^4 or 4^3?

Answer : _____

③ Write 243 as a power of 3.

Answer : _____

④ At Princess Lake there are 100 ducks. If the number of ducks at the lake doubles every 6 months, how many ducks will there be in 2 years?

Answer : _____

⑤ Pat puts 1¢ into her piggy bank on Sep 1, 2¢ on Sep 2, 4¢ on Sep 3, and 8¢ on Sep 4. If this pattern continues, how much will she put into her piggy bank on Sep 10?

Answer : _____

ISBN: 978-1-897164-21-1

Solve the problems and write the answers in scientific notation.

⑥ Express one billion in scientific notation. How many zeros are there?

Answer : _____

⑦ The diameter of the earth is 12 800 km. The diameter of Jupiter is 11 times that of the earth. Calculate the diameter of Jupiter.

Answer : _____

⑧ The distance from the earth to the sun is about 93 million miles.
If 1 mile = 1.6 km, determine the distance in km.

Answer : _____

24 million

= 24 000 000

= 2.4 × 10 000 000

= 2.4×10^7
(in scientific notation)

Read this first.

⑨ The population of Pleasantville is doubling every 20 years. If the current population is 100 000, what will the population be in 40 years?

Answer : _____

⑩ What was the population of Pleasantville 20 years ago?

Answer : _____

CHALLENGE

① A worm crawls along a 100 m log. Each day it crawls half of the remaining distance. After 4 days, how far is it from the end of the log?

Answer : _____

② There are 120 bacteria in a culture that doubles in size every day. How many bacteria were there 3 days ago?

Answer : _____

EXAMPLE

Ruth bought $300.00 worth of Canada Savings Bonds which pay 5% simple interest. What is the value of the bonds after 4 years?

Note : **Interest (I) = Principal (P) × Interest Rate (R) × Time (T)**

Accumulated Amount (A) = Principal (P) + Interest (I)

Total interest after 4 years $= \overset{3}{300} \times \dfrac{5}{\underset{1}{100}} \times 4$

$= 60$

Value of the bonds $=$ Principal $+$ Interest

$= 300 + 60$

$= 360$

Answer : The value of the bonds is $360.00.

Complete the table and solve the problems.

	Name	Principal ($)	Annual simple interest rate	Number of years	Interest gained ($)	Total amount received ($)
①	Clara	400.00	$2\frac{1}{2}$ %	2		
②	Betty	1000.00	5 %	3		
③	James	600.00	4 %	4		
④	Andy	250.00	2 %	3		

⑤ Peggy deposited $1000.00 in a savings account which pays $1\frac{1}{2}$ % simple interest. The account was closed after $1\frac{1}{2}$ years. How much money did Peggy receive?

⑥ Jane invests $2000.00 in a Guaranteed Investment Certificate (GIC) which pays 7% simple interest. How much can she accumulate in 5 years?

Answer : _____

Answer : _____

Solve the problems. Show your work.

⑦ Bank Uno pays $68.50 interest on an investment of $1000.00 for 1 year. Bank Duo pays $6.50 interest on an investment of $100.00 for 1 year. Which bank offers the better rate?

Answer : _____

⑧ Jim bought $8000.00 worth of Canada Savings Bonds and cashed in for $12 000.00 5 years later. What was the simple interest rate paid on these bonds?

Answer : _____

⑨ Kay borrowed $1500.00 from the bank and paid $1800.00 some time later. If the interest rate was 10%, when did she pay back the loan?

Answer : _____

⑩ Bank Alta will loan $2000.00 to Bill if he agrees to pay back $3000.00 two years later. Bank Bee will loan him $2000.00 if he agrees to pay back $3400.00 three years later. Which bank should Bill choose?

Answer : _____

CHALLENGE

The local credit union advertises that you can double your money in 12 years if you invest with them. What simple interest rate are they offering? (correct to the nearest 0.1%)

Answer : _____

EXAMPLE

4 out of 32 students failed a Math test.

a. What percentage of the students failed the test?

$$\frac{\overset{1}{\cancel{4}}}{\underset{2}{\overset{8}{\cancel{32}}}} \times \overset{25}{\cancel{100}}\,\% = \frac{25}{2}\,\% \rightarrow 12\frac{1}{2}\,\%$$

Answer: $12\frac{1}{2}\,\%$ of the students failed the test.

b. What percentage passed?

$$\frac{32-4}{32} \times 100\,\% = \frac{28}{32} \times 100\,\% = \frac{7}{\underset{2}{8}} \times \overset{25}{\cancel{100}}\,\% = \frac{175}{2}\,\% \rightarrow 87\frac{1}{2}\,\%$$

Answer: $87\frac{1}{2}\,\%$ of the students passed the test.

Solve the problems. Show your work.

① At birth a baby is about 30% of its adult height. If Mr Smith is 170 cm tall, estimate his length at birth.

Answer: _____

② Ben's parents give him a 10% raise in his weekly allowance. His new allowance is $22.00. What was his old allowance?

Answer: _____

③ Computers are sold at a discount of 35%. How much must one pay for a computer if the original price was $3200.00?

Answer: _____

④ John has 20 hockey cards and Tim has 64. Tim gives John 25% of his cards.

a. How many cards does John have now?

Answer: _____

b. By what percentage has the number of John's cards increased?

Answer: _____

⑤ Alison gets 12 out of 15 on a quiz. Ruth gets 15 out of 18 on another quiz. Who has a higher score?

Answer: _____

ISBN: 978-1-897164-21-1

⑥ Jo buys a pair of jeans at $59.99, 2 T-shirts at $34.99 each, and 1 pair of shoes at $79.99.

Read this first.

Tax Hint
• A quick way to calculate 15% of tax:

e.g. 15% of $14.60

a. 10% of 14.60 = $1.46
b. 5% of 14.60 = $0.73

Tax : $1.46 + $0.73 = $2.19

a. How much does she pay altogether before tax?

Answer : _____

b. If the tax rate is 15%, how much tax does she pay? (correct to 2 decimal places)

Answer : _____

c. What is the total cost including tax?

Answer : _____

⑦ Susan bought some CDs. She paid $9.00 in tax (tax rate is 15%). How much did she pay for the CDs before tax?

Answer : _____

⑧ In Britain a Value Added Tax (VAT) of 17% is added to most purchases. If a skirt costs $58.50 including tax, what is the price of the skirt before tax?

Answer : _____

⑨ Mrs Jones sold 2 houses for $225 000.00 and $170 000.00 respectively. Calculate her $2\frac{1}{2}$% commission on the house sales.

Answer : _____

⑩ 1 in 40 of the world's bird species is considered to be endangered.

    a.   What percentage of bird species are endangered?

Answer : _____

    b.   If there are 1000 different species of birds in Canada, how many species are likely to be endangered?

Answer : _____

⑪ 40% of North America's national parks are located in Canada. If there are 34 national parks in Canada, how many are there in North America?

Answer : _____

⑫ Julia's boss promises her a 5% raise every 6 months if she stays with the company. Her current salary is $2000.00 per month. What should she expect to earn 1 year from now?

Answer : _____

⑬ Low fat cheese has 30% less calories than regular cheese. If a piece of low fat cheese contains 210 calories, how many calories are there in a piece of regular cheese of the same size?

Answer : _____

⑭ Tim has bought a jacket at $120.00. If he buys it during the Christmas sale, he will pay only $102.00. Calculate the percentage discount.

Answer : _____

ISBN: 978-1-897164-21-1

Use the table below to solve the problems.

⑮ UNICEF collects $3.4 million in its orange Halloween boxes, most of it coins. Here is a breakdown of coins in percentage.

Coin	Penny	Nickel	Dime	Quarter	Loonie	Toonie
Revenue	19%	11%	21%	26%	12%	11%
Coins	77%	9.4%	8.6%	4.2%	0.5%	0.3%

a. Which coins make up the largest contribution to UNICEF?

Answer : _____

b. Assuming that all the money collected is in coins, how much money collected is in dimes?

Answer : _____

c. Explain why the 2 rows in the chart show different percentages.

Answer : _____

CHALLENGE

The chart shows the average annual interest rates for Canadian savings accounts between 1990 and 1993.

Year	1990	1991	1992	1993
Interest rate	8.8%	4.5%	2.2%	0.8%

Judy deposited $1000.00 in a savings account on Jan 1, 1990. How much was in her account on Dec 31, 1993? (correct to 2 decimal places)

Answer : _____

MIDWAY REVIEW

Circle the correct answer in each problem.

Mr Smith has a field in Milton. The field is square in shape and has an area of 22 500 m².

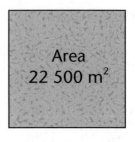

Area
22 500 m²

① If Mr Smith divides the field into 5^4 equal parts. What is the area of each part?

A. 36 m²

B. 21.97 m²

C. 1125 m²

D. None of the above

② Mr Smith wants to enclose the field on all 4 sides with a fence at a cost of $85.00 for every 20 m. How much should Mr Smith pay?

A. $637.50

B. $127.50

C. $51 000.00

D. $2550.00

③ Mr Smith has built a square-shaped pond in the middle of the field, leaving 60 m of land on each side. Which expression shows the area of the pond?

A. $150 - (60 + 60)^2$ m²

B. $(150 - 60)^2$ m²

C. $(150 - 60 - 60)^2$ m²

D. $(150 - 60 + 60)^2$ m²

④ If Mr Smith builds the pond at the corner of the field instead of in the middle, the area of land left behind for cultivation

A. will remain unchanged.

B. will increase by 10%.

C. will increase by 5%.

D. will decrease by 5%.

⑤ Mr Smith wants to apply fertilizer to the soil after building the pond. Each 20-kg bag of fertilizer can treat 100 m² of land. How many kilograms of fertilizer should Mr Smith buy?

A. 4500 kg

B. 4320 kg

C. 4000 kg

D. None of the above

⑥ The fertilizer costs $6480.00. Mr Smith promises to pay the amount with interest in $1\frac{1}{2}$ years. If the simple interest rate is 5%, how much should Mr Smith pay?

A. $6804.00

B. $4860.00

C. $6966.00

D. $7326.00

ISBN: 978-1-897164-21-1

Look at the diagrams and solve the problems.

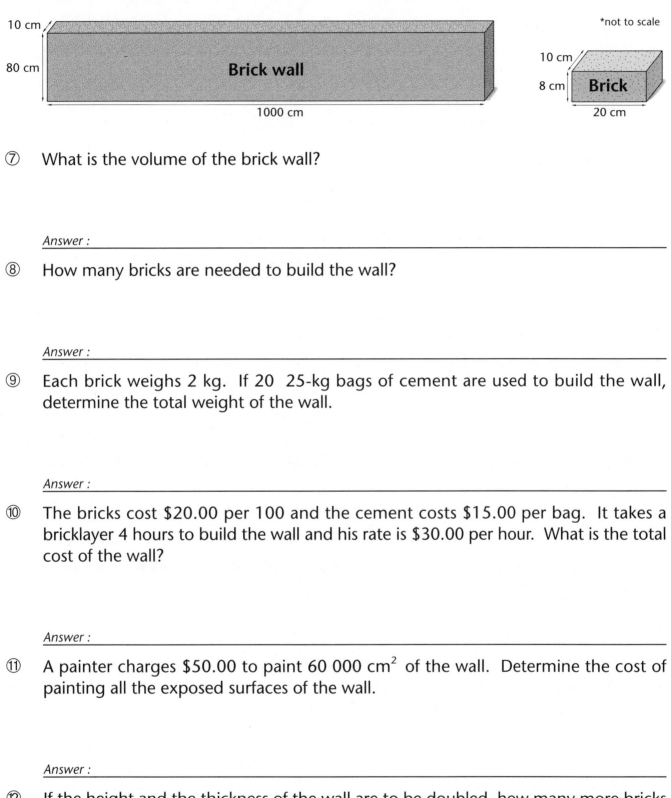

10 cm

80 cm

Brick wall

1000 cm

*not to scale

10 cm

8 cm   Brick

20 cm

⑦ What is the volume of the brick wall?

Answer :

⑧ How many bricks are needed to build the wall?

Answer :

⑨ Each brick weighs 2 kg. If 20 25-kg bags of cement are used to build the wall, determine the total weight of the wall.

Answer :

⑩ The bricks cost $20.00 per 100 and the cement costs $15.00 per bag. It takes a bricklayer 4 hours to build the wall and his rate is $30.00 per hour. What is the total cost of the wall?

Answer :

⑪ A painter charges $50.00 to paint 60 000 cm² of the wall. Determine the cost of painting all the exposed surfaces of the wall.

Answer :

⑫ If the height and the thickness of the wall are to be doubled, how many more bricks will be needed?

Answer :

Look at the map and solve the problems.

Abbeyville

150 km

450 km   Capetown

Beaufort

⑬ Dave travels from Abbeyville to Beaufort in 8 hours. What is his average speed?

Answer : _____

⑭ If Dave sets off from Abbeyville at noon, when will he reach Capetown?

Answer : _____

⑮ Eric travels from Beaufort to Abbeyville in 6 hours. What is his average speed?

Answer : _____

⑯ If Eric sets off from Beaufort at noon, when will he be 50 km beyond Capetown?

Answer : _____

⑰ If Dave and Eric set off from Abbeyville together, how far apart will they be after 3 hours of travelling?

Answer : _____

⑱ Dave sets off from Abbeyville and Eric sets off from Beaufort at the same time. Will they meet after 3 hours of travelling?

Answer : _____

⑲ Jim travels from Beaufort to Abbeyville at an average speed of 30 km/h. If he sets off at 7 a.m., will he reach Capetown by noon?

Answer : _____

Use the table below to solve the problems.

Population by Age Group (1998)					
Province	Canada	Quebec	Ontario	Manitoba	Saskatchewan
Number (correct to 100s)					
All ages	30,301,200	7,334,500	11,413,600	1,138700	1,024,300
Male	14,998,900	3,617,100	5,625,100	564,400	509,500
Female	15,302,300	3,717,400	5,788,500	574,300	514,800
0 - 14	5,975,800	1,357,200	2,282,800	243,800	227,600
Male	3,064,100	693,600	1,171,200	125,000	116,500
Female	2,911,700	663,600	1,111,600	118,800	111,100
15 - 64	20,588,300	5,065,800	7,711,500	739,700	647,700
Male	10,345,600	2,549,600	3,849,300	374,200	328,100
Female	10,242,700	2,516,200	3,862,200	365,500	319,600
65 and over	3,737,100	911,500	1,419,400	155,200	148,900
Male	1,589,200	373,900	604,600	65,300	64,900
Female	2,147,900	537,600	814,800	89,900	84,000

⑳   What percentage of Canadians live in Ontario? (correct to the nearest 0.1%)

Answer : _____

㉑   What percentage of Canadians are female? (correct to the nearest 0.1%)

Answer : _____

㉒   Which one of these provinces - Quebec, Ontario, Manitoba or Saskatchewan - has the youngest population? Explain.

Answer : _____

㉓   Write the population of Canada in scientific notation. (correct to 1 decimal place)

Answer : _____

㉔   Does the percentage of male Canadians remain the same for different ages? Explain your answer.

Answer : _____

EXAMPLE

Janet eats $\frac{2}{3}$ of a 120-g bag of potato chips. John eats $\frac{3}{5}$ of a 125-g bag of chips. Who eats more?

Janet eats $\frac{2}{\underset{1}{3}} \times \overset{40}{120} = 80$ 　　　　　John eats $\frac{3}{\underset{1}{5}} \times \overset{25}{125} = 75$

Answer: Janet eats more.

* Note : Always write fractions in simplest form, e.g. $\frac{18}{21} = \frac{6}{7}$

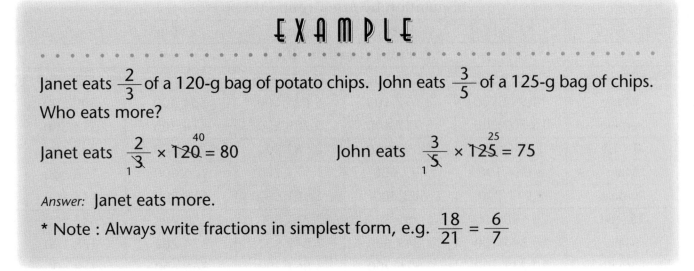

Colour and write.

① Colour the diagrams and write the mathematical sentences.

a.

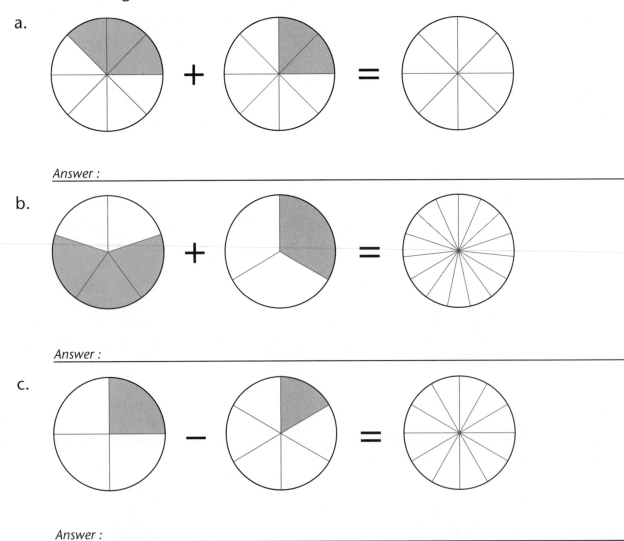

Answer : _____

b.

Answer : _____

c.

Answer : _____

ISBN: 978-1-897164-21-1

Solve the problems. Write the answers in simplest form.

② Daniel spent $2\frac{1}{2}$ hours playing computer games on Saturday and $1\frac{1}{3}$ hours on Sunday.

    a. How much time did he spend playing computer games altogether that weekend?

Answer : _____

    b. The following weekend, he spent only $2\frac{3}{4}$ hours playing computer games. How much less time did he spend compared with the previous weekend?

Answer : _____

③ A pizza costs $7.95. How much would you pay for $\frac{5}{12}$ of the pizza? (correct to the nearest cent)

Answer : _____

④ Judy ate $\frac{3}{4}$ of a 240 g chocolate bar and $\frac{2}{3}$ of a 210 g bar. How much chocolate did she eat?

Answer : _____

⑤ A bag of 12 apples weighs 1500 g. How much would $\frac{2}{3}$ of the bag of apples weigh? What are you assuming?

Answer : _____

⑥ Bill cut a cake into 14 pieces and ate 3 of them. Carol cut a same size cake into 16 pieces and ate 5 of them. Who ate more?

Answer : _____

⑦ It takes 3 minutes to half fill a bath. How long would it take to fill $\frac{3}{4}$ of the bath at the same rate?

Answer : _____

ISBN: 978-1-897164-21-1

⑧ Adam gets 10 out of 14 questions correct on a quiz. Bert gets 12 out of 16 questions correct on another quiz. If all questions carry equal marks, who gets the better grade?

Answer :
_____

⑨ In a parking lot, $\frac{1}{3}$ of the cars are white, $\frac{1}{4}$ black and $\frac{1}{5}$ red. What fraction of the cars are neither white, black nor red?

Answer :
_____

⑩ Janet wants to build a 12 m fencing, using $1\frac{1}{4}$ m long planks. How many planks must she buy?

Answer :
_____

⑪ Mary has bought $13\frac{1}{2}$ m of fabric. She needs 3 m to make a dress. How many dresses can she make?

Answer :
_____

⑫ A reading room is half full. It becomes $\frac{1}{3}$ full when 10 people have left the room. Determine how many people the room holds when full.

Answer :
_____

⑬ Mrs Black wants to put some textbooks on a 90 cm long shelf. If each book is $4\frac{1}{4}$ cm thick, how many textbooks will fit on the shelf?

Answer :
_____

⑭ What is the average of $\frac{1}{4}$ and $\frac{3}{4}$?

Answer :
_____

     ISBN: 978-1-897164-21-1

Use the table to solve the problems.

⑮ The table shows the calories in tomato juice.

a. Complete the table. (round to the nearest whole number)

Serving	$\frac{1}{2}$	$\frac{2}{3}$	$\frac{3}{4}$	1	$1\frac{1}{4}$	$1\frac{1}{2}$
Calorie			105			

b. How many servings would give 350 calories? (write the answer in fraction)

Answer : _____

⑯ The table shows the population of Hicksville from 1970 to 2000.

Year	1970	1980	1990	2000
Population (to the nearest ten thousand)	2	$2\frac{17}{20}$	$2\frac{1}{2}$	$2\frac{3}{10}$

a. When was the fastest population growth?

Answer : _____

b. How many more people lived in Hicksville in 1980 than in 1970?

Answer : _____

c. Find the percentage change of population in Hickville from 1980 to 2000. Does the population increase or decrease? Use the percentage change to explain. (correct to the nearest 0.1%)

Answer : _____

$$Percentage\ change = \frac{New\ Value - Original\ Value}{Original\ Value} \times 100\%$$

$Percentage\ change$ $\begin{cases} > 0 & percent\ increase & : Original\ Value < New\ Value \\ = 0 & no\ change & : Original\ Value = New\ Value \\ < 0 & percent\ decrease & : Original\ Value > New\ Value \end{cases}$

Read this first.

⑰ Read the facts about the Grade 7 classes in Riverdale Junior High School and solve the following questions.

Fraction of students in Grade 7		$\frac{2}{17}$
Number of students	Class A	30 ($\frac{1}{2}$ are boys)
	Class B	36 ($\frac{2}{3}$ are boys)
Average time spent on homework per day		English: $\frac{1}{5}$h ; Math: $\frac{1}{3}$h Science: $\frac{1}{4}$h ; Social Science: $\frac{1}{4}$h
Number of students who scored 80 marks or more on Math test	Boy	20
	Girl	15

a. If there are 66 students in Grade 7, how many students are there in the school?

Answer : _____

b. What fraction of the Grade 7 students are girls?

Answer : _____

c. How many Grade 7 students are boys?

Answer : _____

d. What percentage of the Grade 7 students are boys? (correct to the nearest 0.1%)

Answer : _____

e. How much time does a Grade 7 student spend on homework each day?

Answer : _____

f. How much time does a Grade 7 student spend on homework during weekdays?

Answer : _____

g. What percentage of the Grade 7 students scored 80 marks or more on the Math test? (correct to the nearest 0.1%)

Answer : _____

ISBN: 978-1-897164-21-1

Solve the problems. Show your work.

⑱ Andrew and Anna were reading a 320-page novel. On the first day, both of them read $\frac{1}{4}$ of the book. On the second day, Andrew read $\frac{1}{8}$ of the book while Anna read $\frac{1}{8}$ of the remaining pages.

a. How many pages does Andrew still have to read?

Answer : _____

b. How many pages does Anna still have to read?

Answer : _____

c. Who read more pages on the second day? How many pages more?

Answer : _____

⑲ On the Toronto Stock Exchange (TSE), the Richmond Bank Stock was listed at $\$37\frac{1}{2}$ on Monday and $\$32\frac{3}{4}$ on Tuesday.

a. What is the change in the stock price of Richmond Bank?

Answer : _____

b. If this trend continues, what would the stock price be on Wednesday?

Answer : _____

CHALLENGE

You own a stock which increased in price by $\$1\frac{3}{8}$, then decreased by $\$\frac{5}{8}$, then increased by $\$\frac{1}{4}$ and then decreased by $\$\frac{7}{8}$. Is your stock worth more or less than at first?

Answer : _____

ISBN: 978-1-897164-21-1

UNIT 8 Decimals

EXAMPLE

Estimate and then calculate the total cost of the following items:

2 pairs of jeans at $59.25 each, 1 jacket at $119.95, 3 T-shirts at $29.45 each and 2 pairs of shorts at $34.99 each.

Estimate			Calculation		
2 × 60	=	120	2 × 59.25	=	118.50
1 × 120	=	120	1 × 119.95	=	119.95
3 × 30	=	90	3 × 29.45	=	88.35
2 × 35	= +	70	2 × 34.99	= +	69.98
		400			396.78

Answer : The total cost is $396.78.

Solve the problems. Write the answers correct to 2 decimal places.

① John exchanged US$150.00 into Canadian dollars on a day when the exchange rate was C$1.00 = US$0.68. How much money did he get?

Answer : John got C$ _____

② Jim exchanged C$78.00 into US dollars on a day when the exchange rate was C$1.00 = US$0.68. How much money did he get?

Answer : _____

③ Determine the better value for each of the following items. (Use C$1.00 = US$0.68) Check ✓ the correct answers.

	Item	US($)		Canadian ($)	
a.	1 pair of jeans	49.99	◯	69.99	◯
b.	1 CD	14.99	◯	22.75	◯
c.	Pop	6.99 for 24 cans	◯	4.99 for 12 cans	◯
d.	1 T-shirt	25.99 + 5% tax	◯	35.99 + 15% tax	◯

ISBN: 978-1-897164-21-1

④ You have $12.50. How many comic books can you buy if they cost $2.54 each?

Answer : _____

⑤ Bill commutes 320 km each week to and from work. His gas consumption averages 7 L per 100 kilometres and gas costs $0.63/L. Calculate his weekly expenditure on gas.

Answer : _____

⑥ Derek got $100.00 for his birthday. He bought 5 CDs which cost $15.49 each. How much money did he have left for lunch?

Answer : _____

⑦ John paid $150.00 to buy 3 CDs at $16.99 each and 2 T-shirts at $24.25 each, plus 15% tax.

a. How much did he spend?

Answer : _____

b. How much change did he get?

Answer : _____

⑧ Tony and 7 of his friends wanted to share $4.75 equally. How much money would each of them get?

Answer : _____

⑨ Pat has $3.48. What is the maximum number of nickels she has?

Answer : _____

⑩ A certain map has a scale of 150 km to 1 cm. If 2 towns are 3.73 cm apart on the map, how far are they on land?

Answer : _____

⑪ The length of an Earth Year is 365.3 days. If it takes Mars 1.88 times as long as the earth to circle the sun once, determine the length of a Mars Year.

Answer : _____

⑫ The first woman in space made 48 orbits around the earth in 70.83 hours.

a. How long did each orbit take?

Answer : _____

b. How many orbits did she make in 10 hours?

Answer : _____

⑬ When a number is added $\frac{1}{4}$ of itself, the result is 4.5. Find the number.

Answer : _____

Use the table below to solve the problems. Write the answers correct to 3 decimal places.

⑭ Calculate the batting average of each batter. Complete the table.

Lightning Baseball Team - Performance of Batters			
Period : June 1, 2000 to July 10, 2000			
Name	Times at bat	Number of hits	Batting average
J. Green	14	5	5 ÷ 14 =
S. White	13	4	
C. Brown	12	4	
K. Jones	16	5	

⑮ J. Green was at bat 3 times without a hit in the game on July 11, 2000. What was his new batting average?

Answer : _____

⑯ The batting average of S. White rose to 0.375 after the match on July 11, 2000. He was at bat 3 times in the game. How many more hits did he get?

Answer : _____

⑰ The batting average of C. Brown remained the same though he got 1 hit in the game on July 11, 2000. How many times was he at bat that day?

Answer : _____

⑱ What was the batting average of the four batters on July 10, 2000?

Answer : _____

   ISBN: 978-1-897164-21-1

Use the map below to solve the problems. Write the answers correct to 3 decimal places.

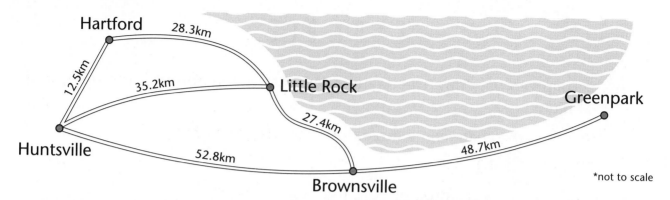

⑲ Tony drives from Huntsville to Greenpark via Little Rock and Brownsville at a speed of 65.5 km/h every day. How long does he take to reach Greenpark?

Answer : _____

⑳ Mr Wong takes 2.2 hours driving from Huntsville to Greenpark via Hartford, Little Rock and Brownsville every day. What is his average speed?

Answer : _____

㉑ Mr Black took 2.90 hours to travel from Huntsville to Greenpark and back. He kept an average speed of 70 km/h. Which route did he take?

Answer : _____

Elaine has saved $15.35 in loonies, dimes and nickels. 19 nickels are collected and there are 2 times as many dimes as loonies. How many dimes and loonies has Elaine saved?

Answer : _____

ISBN: 978-1-897164-21-1

EXAMPLE

On a certain February day, the daytime temperature in Toronto was -12°C and -22°C at night.

a. What was the change in temperature from day to night?

-22 – (-12) = -10

Answer : There was a drop of 10°C from day to night.

b. What was the average temperature over the day and night?

(-22 + (-12)) ÷ 2 = (-22 – 12) ÷ 2 = (-34) ÷2 = -17

Answer : The average temperature over the day and night was -17°C.

Complete each of the statements with 'always', 'sometimes' or 'never'.

① The sum of 2 positive integers is _____ positive.

② The sum of a positive integer and a negative integer is _____ negative.

③ The sum of 2 negative integers is _____ positive.

④ A positive integer is _____ smaller than a negative integer.

⑤ If 2 integers are negative, then the one closer to zero is _____ greater.

Solve the problems. Show your work.

⑥ What is the sum of two opposite numbers? Explain your answer with a number line.

• Opposite numbers are two numbers that are the same distance from zero but in opposite directions on a number line, e.g. +5 and –5.

Read this first.

Answer : _____

ISBN: 978-1-897164-21-1

⑦ A vulture can fly at a height of 8000 m. A goose can only fly at 1000 m. A turtle can dive to a depth of 1200 m.

a. Represent these values as integers on the diagram.

m

| 8000 |
| 7000 |
| 6000 |
| 5000 |
| 4000 |
| 3000 |
| 2000 |
| 1000 |
| 0 | sea level |
| -1000 |
| -2000 |

b. Determine the sum of these integers.

Answer : _____

c. What is the difference between the heights of the vulture and the goose?

Answer : _____

d. What is the difference between the heights of the vulture and the turtle?

Answer : _____

⑧ Mount Logan in Yukon is 5900 m high. A trench under the Pacific Ocean is 11 000 m below sea level.

a. Write each of the measurements as integers.

Answer : _____

b. Would Mount Logan be seen as an island if it rose from the floor of the trench? Explain with an integer-expression.

Answer : _____

⑨ The initial price of MBank Stock was 23\frac{1}{8}$. The following changes were recorded from Monday to Friday last week: down 2\frac{1}{4}$, up 2\frac{3}{4}$, down 3\frac{1}{2}$, up $2 and down 3\frac{3}{4}$.

a. Write an integer-expression to show the changes.

Answer : _____

b. Mr Smith bought 1000 units of MBank shares last Monday and sold all of them on Friday. How much did he gain or lose?

Answer : _____

ISBN: 978-1-897164-21-1

⑩ The minimum surface temperatures of some planets are given in the chart.

a. Which planet is the coldest?

Answer : _____

Planet	Minimum temperature
Mercury	-184°C
Earth	-90°C
Mars	-123°C

b. What is the difference between the minimum temperatures of the earth and Mars?

c. List the temperatures in order using > or < signs.

Answer : _____

Answer : _____

⑪ The sum of temperatures in Winnipeg and Calgary on a cold winter day was -35°C. If the temperature in Winnipeg was -19°C,

a. what was the temperature in Calgary?

Answer : _____

b. what was the difference between the temperatures?

Answer : _____

⑫ Study the graph which shows the temperatures in Toronto from January 1 to January 7, 1999.

a. Write an integer-expression to show the change in daily temperature from January 1 to January 7.

Answer : _____

b. If the temperature on January 8 was down 7°C, what was the temperature that day?

Answer : _____

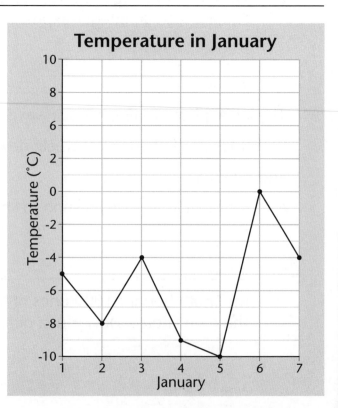

Temperature in January

ISBN: 978-1-897164-21-1

⑬ In a certain football game the running back gained 6 m, then lost 3 m, then lost 2 m, then lost 4 m. Write the numbers as integers and determine whether he had gained or lost ground overall.

Answer : _____

⑭ Mr Smith is a golfer. The chart below shows his performance in a 9-hole tournament.

Hole								
1	2	3	4	5	6	7	8	9
par	double bogey	birdie	eagle	par	bogey	bogey	birdie	double bogey

par : the number of strokes one should take to hit the ball into a hole
double bogey : 2 over par   eagle : 2 under par
bogey   : 1 over par   birdie : 1 under par

a. Was Mr Smith's score over or under par?

Answer : _____

b. If par for the 9 holes is 36, what was his score?

Answer : _____

CHALLENGE

① Annie is thinking of 2 integers. Their sum is -9 and their difference is 5. What are the integers?

Answer : _____

② Find two integers with an average of 9 and a difference of 8.

Answer : _____

£XAMPL£

Jill put 1¢ into her piggy bank on the first day, 2¢ the second day, 4¢ the third day, etc.

a.   How much did she deposit on the seventh day?

Sequence:

Day	1	2	3	4	5	6	7
Deposit (¢)	1	2	4	8	16	32	64

×2   ×2   ×2   ×2   ×2   ×2

Answer : She deposited 64¢ on the seventh day.

b.   How much did she save altogether over the first 7 days?

Total savings :   1 + 2 + 4 + 8 + 16 + 32 + 64 = 127

Answer : She saved 127¢ ($1.27) over the first seven days.

Write the pattern for each of the problems. Show your work.

①   The cost of a taxi ride includes an initial charge of $2.00. If a 5 km ride costs $6.00 and a 10 km ride costs $10.00, determine:

a.   the cost of a 15 km ride.

Answer : _____

b.   the cost of a 40 km ride.

Answer : _____

c.   how far you can travel for $18.00.

Answer : _____

ISBN: 978-1-897164-21-1

② A baby chick weighs 100 g at birth. If it doubles its weight each week, how many weeks will it take to reach a weight of at least 3 kg?

Answer : _____

③ There were 500 students at Richview School in 1960, 450 in 1970, 550 in 1980 and 500 in 1990. If this trend continues, how many students will attend Richview School in 2020?

Answer : _____

④ A baseball card is bought for $2.00. If it increases in value by $0.50 each week,

a. when will it be worth $6.00?

Answer : _____

b. how much will it be worth after 10 weeks?

Answer : _____

⑤ Adam and David were growing a mold for a science project. It weighed 5 g on the first day, 9 g the second day, 17 g the third day and 33 g the fourth day.

a. Describe the pattern you observe.

Answer : _____

b. If this pattern continues, how much will it weigh on the seventh day?

Answer : _____

⑥ James sold a precious stamp for $45.00. If its value increased by $3.00 each of the 6 years he owned it, what was the original price?

Answer : _____

⑦ Mrs Jones monitored the number of books on a certain library shelf. There were 52 books on the shelf the first day, 48 the second day, 50 the third day, 46 the fourth day and 48 the fifth day.

a.   If this pattern continues, how many books will there be on the tenth day?

Answer : _____

b.   When will there be 36 books remaining on the shelf?

Answer : _____

⑧ The first number of a pattern is 3 and the sixth number is 96. The pattern is created using multiplication.

a.   Write the missing numbers in the pattern.

Answer : 3,   ,   ,   ,   , 96 _____

b.   What is the eighth number?

Answer : _____

⑨ The following pattern was discovered by an Italian mathematician called Fibonacci. Describe the pattern and find the next 5 terms.

$$1, \quad 1, \quad 2, \quad 3, \quad 5, \quad 8, \quad 13$$

Answer : _____

⑩ There were exactly 4 Wednesdays and 4 Saturdays in September one year. On what day did September 1 fall that year? Explain.

Answer : _____

ISBN: 978-1-897164-21-1

⑪ The numbers 1, 4, 9, 16 ... form a pattern.

a. Describe the pattern.

Answer : _____

b. What is the sixth number in the pattern?

Answer : _____

⑫

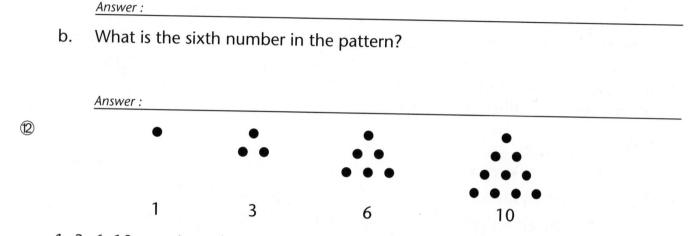

1   3   6   10

1, 3, 6, 10 are triangular numbers. Find the seventh triangular number.

Answer : _____

⑬ Each number in the sequence in question 11 can be written as the sum of a series of consecutive odd numbers beginning with 1, e.g. $16 = 1 + 3 + 5 + 7$.

a. Write the sixth number as the sum of a series of consecutive odd numbers. How many numbers are there in the series?

Answer : _____

b. Write the tenth number as the sum of a series of consecutive odd numbers and determine its value.

Answer : _____

CHALLENGE

The n th triangular number is determined to be the sum of the first n positive whole numbers, e.g. $6 = 1 + 2 + 3$. Therefore 6 is the third triangular number. If the 100th triangular number is 5050, determine the 101st triangular number.

Answer : _____

ISBN: 978-1-897164-21-1

11 Linear Equations

EXAMPLE

Peter has a 2-pan balance. He stacks a pile of 10-g weights in one pan and puts a 300-g weight in the other. To balance the pans, he needs to add a 50-g weight to the 10-g weights. Write an equation to represent this situation and solve it to determine the number of 10-g weights.

Solution:

A. **Define the variable** ➤ Let n be the number of 10-g weights

B. **Write an equation** ➤ $10n + 50 = 300$

C. **Solve the equation** ➤
$$10n + 50 - 50 = 300 - 50$$
$$10n = 250$$
$$10n \div 10 = 250 \div 10$$
$$n = 25$$

D. **Write the answer in words** ➤ He uses 25 10-g weights.

Write an equation for each of the problems and solve the equations.

① In a diving contest, the score times the degree of difficulty gives the number of points awarded. A diver got 86.08 points for a dive of 3.2 in degree of difficulty. What was his score?

Answer : _____

② St. John's has about 10 cm less than 5 times the snowfall of Vancouver. In 1990 St. John's snowfall was 360 cm. What was the snowfall in Vancouver?

Answer : _____

③ Find 2 numbers with a sum of 40 and a difference of 2.

Answer : _____

 ISBN: 978-1-897164-21-1

④ A plumber charges $30.00 per call plus $25.00 per hour. If a certain repair costs $80.00, how long does it take?

Answer : _____

⑤ 2 sides of a triangle measure 6 cm and 13 cm. The perimeter of the triangle is 28 cm. What is the length of the third side?

Answer : _____

⑥ The Kim family rented a car for a day trip. Mr Kim paid $30.00 per day plus $0.10 per km. If the total cost was $50.00, how far did the Kims travel?

Answer : _____

⑦ A letter carrier walks about 6.8 km daily, which is about 1.2 km more than double the distance walked by a doctor. How far does a doctor walk?

Answer : _____

⑧ The Bellvue Health Club charges each member $24.00 per month plus $2.00 per exercise class. Anna spends $40.00 per month. How many exercise classes does she attend?

Answer : _____

⑨ Carol thinks of a number. She multiplies it by 3 and then subtracts 15. The answer is 21. What is the number?

Answer : _____

⑩ Mr Mark paid a monthly utility bill of $52.00, which included a late fee of $6.00. If electricity costs $0.20 per kWH, how many kWH of electricity did Mr Mark consume that month?

Answer : _____

⑪ A local utility company is offering a rebate of $20.00 to families who install efficient light bulbs. The Jones family spent $50.00 on efficient light bulbs last month and saved $1.50 on their utility bill. How long will it take for the Jones family to recover the additional cost?

Answer : _____

⑫ There are 83 Grade 7 students in Queenston Junior High School. The number of boys is 31 less than 2 times the number of girls. How many boys and girls are there?

Answer : _____

⑬ David subtracted 15 from a number. He then doubled the answer to get a final value of 22. What is the original number?

Answer : _____

⑭ A box of 24 tea bags costs $1.70 to produce, including $0.50 of packaging cost.

　　a.　How much does each tea bag cost?

Answer : _____

　　b.　If the packaging cost remains the same, how much does it cost to produce a 75-bag box?

Answer : _____

ISBN: 978-1-897164-21-1

⑮ A dining room chandelier has a maximum wattage of 600 W. If there are 8 identical bulbs surrounding a 120 W central bulb, what is the maximum allowable wattage of each of the surrounding bulbs?

Answer :
_____

⑯ A deep sea diver dives to a depth of 80 m. He then rises at a rate of 2 m/s. How long would he take to reach a depth of 14 m?

Answer :
_____

⑰ James has saved all his pennies in a box. The box weighs 520 g and each penny weighs 1.2 g. How many pennies has James saved if the total weight of the box and the pennies is 1636 g?

Answer :
_____

⑱ Cindy works in a supermarket. She earns $6.80 per hour plus $20.00 bonus per month. How many hours did she work if she earned $802.00 last month?

Answer :
_____

CHALLENGE

Heating with gas costs about $\frac{2}{3}$ of the cost of heating with electricity. The Wong family spend $120.00 monthly on electric heat, so they plan to spend $600.00 to convert to gas. When will they recover the cost?

Answer :
_____

ISBN: 978-1-897164-21-1

Solve the problems. Show your work.

At Richview Junior High School, there are 70 students in Grade 7, 40 of them boys and 30 girls. 56 of the Grade 7 students are 12 years old and the rest are 13 years old. $\frac{1}{8}$ of the boys and $\frac{1}{3}$ of the girls have joined the Environment Club.

① What fraction of the Grade 7 students are boys?

Answer : _____

② What fraction of the Grade 7 students are 13 years old?

Answer : _____

③ How many students in Grade 7 have joined the Environment Club?

Answer : _____

④ What fraction of the Grade 7 students have joined the Environment Club?

Answer : _____

⑤ What fraction of the club members are girls?

Answer : _____

⑥ How does the fraction of girls in the club compare with the fraction of girls in Grade 7? Discuss.

Answer : _____

⑦ Each member has to pay a membership fee of $1.50. All the male members have paid and $19.50 has been collected. How many female members have paid?

Answer : _____

   ISBN: 978-1-897164-21-1

The Environment Club is planning to organize a trip to the rainforest in Costa Rica. They decide to start fundraising by selling scented candles for the holiday season.

⑧ Complete the chart if the initial pattern continues.

	Day	Total number of candles sold by girls	Total number of candles sold by boys
Start of campaign ▶	0	0	9
	1	5	12
	2	10	15
	3		
	4		
	5		

⑨ If x represents the number of days since the start of the campaign, write expressions for:

a. the number of candles sold by girls.

Answer : _____

b. the number of candles sold by boys.

Answer : _____

c. the number of candles sold by the club members.

Answer : _____

⑩ How many candles will be sold altogether if the campaign continues for 10 days?

Answer : _____

⑪ If the club made a profit of $0.80 per candle sold, calculate how much money would be raised in 10 days?

Answer : _____

⑫   The Environment Club is going to organize pop can recycling in the school. There are 240 students in the school and each drinks an average of 1.2 cans of pop per week. $\frac{3}{4}$ of the cans can be recycled and will be sold to a recycling company for $6.00 per 100 cans.

a.   How many recyclable cans will be collected each week?

Answer : _____

b.   How much money will the club raise in a week?

Answer : _____

⑬   At the end of the term, the Environment Club had a pizza party. 12 club members attended. Each large pizza cost $16.99 + 15% tax and served 6 people. How much did each member pay? (correct to 2 decimal places)

Answer : _____

Use the chart below to solve the problems.

City \ Month	January	April	July	October
Average Daily Temperature in °C of 10 Canadian Cities				
Banff	-7	8	22	10
Calgary	-6	9	23	12
Edmonton	-17	9	22	11
Goose Bay	-12	3	21	7
Halifax	-1	9	20	14
Inuvik	-26	-8	19	-5
Ottawa	-6	11	26	13
Whitehorse	-16	6	20	4
Winnipeg	-14	9	25	12
Yellowknife	-25	-1	21	1

⑭ Which is the hottest city?

Answer :

⑮ Which is the coldest city?

Answer :

⑯ Which of the cities has the most extreme weather conditions?

Answer :

⑰ Which 4 cities have an annual range of temperature over 38°C?

Answer :

⑱ What is the annual range of temperature in Halifax?

Answer :

⑲ Calculate the average temperature in Banff over the four seasons.

Answer :

⑳ Does Yellowknife have an average temperature over freezing or below freezing?

Answer :

㉑ What is the change in temperature from April to October in Inuvik?

Answer :

㉒ Calculate the average temperature of the 10 cities in:

a. January

Answer :

b. October

Answer :

Circle the correct answer in each problem.

㉓ Jim and Alan are removing some boards from a truck. If Jim removes $\frac{1}{8}$ and Alan removes $\frac{1}{4}$ of the boards, there will be 40 boards left. How many boards are there?

A. 50

B. 56

C. 64

D. 78

㉔ Aggie has $24.00 in loonies and toonies. She has 17 coins altogether. How many are toonies?

A. 8

B. 7

C. 6

D. 5

㉕ Half of a number plus $\frac{2}{3}$ of it is 42. The number is:

A. 18

B. 21

C. 28

D. 36

㉖ Gingerale costs $0.79 for 2 cans. Nancy has a $5 bill and wants to buy 1 dozen cans of gingerale. How much change will she get?

A. $0.26

B. $1.26

C. $4.31

D. $4.21

㉗ Determine the next number in the sequence 6, 8, 14, 16, 22 ...

A. 23

B. 24

C. 30

D. 28

㉘ Maggie earns $60.00 per day. She also earns $5.50 per hour in tips. If Maggie worked x hours on Wednesday, the expression for her earnings that day should be:

A. $60 + \dfrac{x}{5.50}$

B. $60 + \dfrac{5.50}{x}$

C. $60 + 5.50x$

D. $60x + 5.50$

㉙ Refer to question 28. How many hours did Maggie work on Wednesday if she earned a total of $104.00?

A. 10

B. 8

C. 9

D. 6

㉚ During the first hour of a party, $\frac{3}{4}$ of the punch was gone. Jane added 5 L to the remaining punch and there was 14 L of punch. If there was x L of punch initially, the equation to represent the problem should be:

A. $\dfrac{1}{4}x + 5 = 14$

B. $\dfrac{1}{4}x - 5 = 14$

C. $\dfrac{3}{4}x + 5 = 14$

D. $\dfrac{3}{4}x - 14 = 5$

㉛ Refer to question 30. The initial amount of punch in litres was:

A. 26

B. 76

C. 36

D. 12

ISBN: 978-1-897164-21-1

Section IV

Overview

Section III includes problems in the Number Sense and Numeration as well as Patterning and Algebra strands.

The focus in Section IV is on word problems that cover the Measurement, Geometry and Spatial Sense, as well as Data Management and Probability.

In Measurement, students will learn more on perimeter, area, volume, and surface area.

In Geometry and Spatial Sense, students will examine congruency and similarity, as well as working with angles and various types of transformations.

The Data Management and Probability strand involves reading, interpreting, and constructing different types of graphs. Students will analyze data through calculating the mean, median, and mode. They will also use tree diagrams to help solve probability problems.

Perimeter and Area

EXAMPLE

A backyard deck has the dimensions shown.

a. What length of fencing is needed to enclose the deck?

b. What area of wooden tiles will cover the deck?

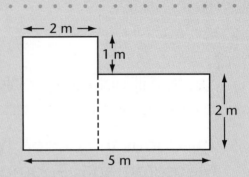

Solutions:

a. Perimeter : $2 + 1 + (5 - 2) + 2 + 5 + (2+1)$

             $= 2 + 1 + 3 + 2 + 5 + 3$

             $= 16$

Answer : 16 m of fencing is needed to enclose the deck.

b. Area : $2 \times (1 + 2) + (5 - 2) \times 2$

         $= 2 \times 3 + 3 \times 2$

         $= 12$

Answer : 12 m² of wooden tiles will cover the deck.

Solve the problems. Show your work.

① A flag is made up of 2 red rectangles with a white square in between containing a red star. Determine the area of each of the red rectangles in the flag shown.

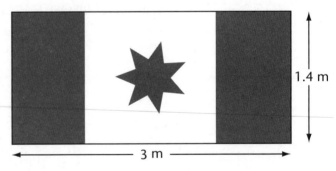

Answer : _____

② 8 circles each having a diameter of 50 cm are drawn in a rectangle as shown. Calculate the area of the rectangle.

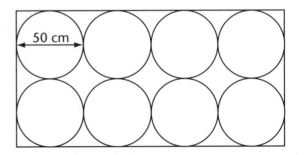

Answer : _____

ISBN: 978-1-897164-21-1

③ A walking group decides to do their 3-km walk around the base of the Toronto Dominion Bank Tower which measures 60 m by 36 m. How many times must they walk around the base of the tower in order to cover at least 3 km?

Answer : _____

④ A rectangular playground is to be constructed with an area of 16 m^2.

a. If the measurement is correct to the nearest 1 m, determine the different possible lengths of fence for enclosing the playground. (Remember a square is also a rectangle.)

Answer : _____

b. What is the minimum length of fencing you could use?

Answer : _____

⑤ The fence around a rectangular field measures 20 m.

a. If the measurement of each side is rounded to the nearest 1 m, determine the possible values for the area of the field.

Answer : _____

b. What is the difference between the maximum and minimum possible areas?

Answer : _____

c. What conclusion can you draw?

Answer : _____

⑥ An artist wants to make a mosaic picture with some small tiles. If the area of each tile is 1 cm^2 and the area of the picture is 1 m^2, how many tiles does the artist need?

Answer : _____

Study the diagrams carefully and solve the problems. Show your work.

Adam and Nadine have bought a house. The diagrams below show the dimensions of the backyard and the living room of their house.

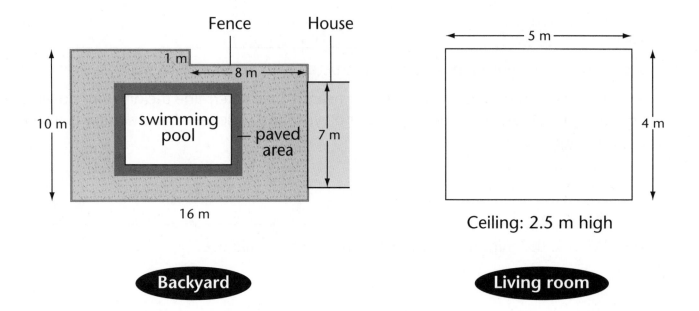

Backyard

Living room

⑦ The backyard is enclosed by fencing which costs $15 /m. How much does the fencing cost?

Answer : _____

⑧ The swimming pool has dimensions of 5 m by 7 m. The paved area is 1 m wide. What is the area of the paved area?

Answer : _____

⑨ The backyard is planted with grass sod which costs $12 /m². What will the cost of the grass be?

⑩ Adam and Nadine want to redecorate their living room. If the whole floor is to be carpeted and the carpeting costs $70 /m² including installation, determine the carpet cost.

Answer : _____

⑪ The walls and the ceiling of the living room are to be painted. The paint costs $8.99 /L and 1 litre covers 5 m². If the windows occupy 7 m², calculate the cost of the paint. (correct to the nearest cent)

Answer : _____

Answer : _____

   ISBN: 978-1-897164-21-1

Solve the problems. Show your work.

⑫ The design on some traditional Canadian quilts is made up of parallelograms as shown. Determine the total area of the design.

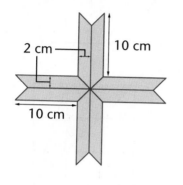

2 cm
10 cm
10 cm

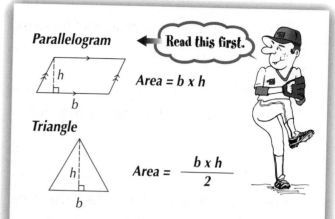

Parallelogram   **Read this first.**

Area = b x h

Triangle

Area = $\dfrac{b \times h}{2}$

- Congruent triangles have the same shape and size.
- An isosceles triangle has 2 sides equal.
- An equilateral triangle has 3 sides equal.

Answer : _____

⑬ 2 congruent equilateral triangles form parallelogram ABCD. The perimeter of each triangle is 24 cm. Determine the perimeter of the parallelogram.

Answer : _____

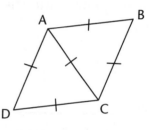

A   B

D   C

⑭ Determine the average area of the three triangles drawn inside the given rectangle with dimensions of 9 cm × 4 cm.

Answer : _____

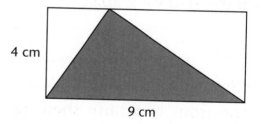

4 cm

9 cm

⑮ A new corporate logo contains an isosceles triangle with sides 6 cm and 12 cm.

a. Determine the perimeter of the triangle.

Answer : _____

b. If the height of the triangle is 11.62 cm, determine its area.

- The sum of any 2 sides of a triangle must be longer than the third side. **Read this first.**

Answer : _____

ISBN: 978-1-897164-21-1

⑯ A hexagonal road sign has the dimensions shown.

a. Determine the perimeter of the sign.

Answer : _____

b. Determine the area of the sign.

Answer : _____

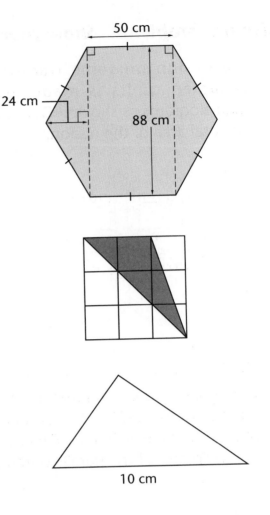

⑰ Determine the shaded area if the area of each square is 1 cm².

Answer : _____

⑱ A triangle with the base of 10 cm has the same area as a square with sides of 5 cm. Determine the height of the triangle.

Answer : _____

⑲ Determine the area and the perimeter of the irregular shape.

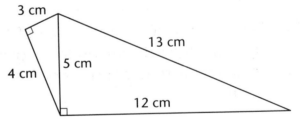

Answer : _____

⑳ The front of a farm shed has the dimensions shown. Determine the area.

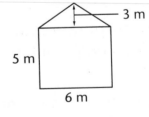

Answer : _____

㉑ The front and back of a blue box are trapezoidal in shape. Determine their total area.

36 cm
30 cm
30 cm

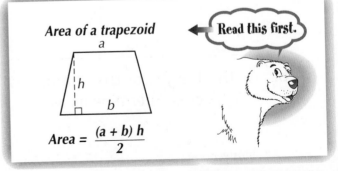

Area of a trapezoid   **Read this first.**

a

h

b

$$Area = \frac{(a + b)\, h}{2}$$

Answer : _____

ISBN: 978-1-897164-21-1

㉒ The sides of the blue box are also trapezoids. Determine their total area.

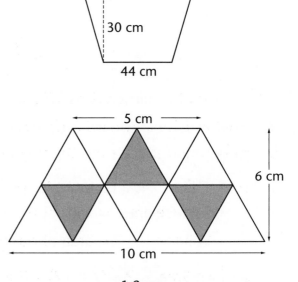

48 cm
30 cm
44 cm

Answer : _____

㉓ The given pattern is part of a quilt design made up of congruent triangles. Determine the total area of the shaded triangles.

5 cm
6 cm
10 cm

Answer : _____

㉔ A carpet contains the hexagonal design shown. Determine the area of the shaded hexagon.

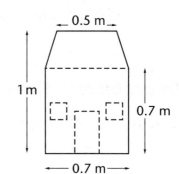

1.2 cm
2 cm
2 cm

Answer : _____

㉕ Suzie's doll house has dimensions as shown. Determine its cross-sectional area.

0.5 m
1 m
0.7 m
0.7 m

Answer : _____

CHALLENGE

A pentomino is a shape made up of 5 identical squares joined at their edges. Draw as many pentomino shapes as you can. Which pentomino has the smallest perimeter?

Answer : _____

Volume and Surface Area

EXAMPLE

A tissue box has dimensions 12 cm by 22 cm by 7 cm.

a. How much cardboard is needed to make the box?

b. How much space does the box occupy?

Think : The box is a rectangular prism.

Surface area = sum of the areas of 6 surfaces (2ab + 2bh + 2ah)

Volume = area of base × height (a × b × h)

a. Surface area : $2 \times (7 \times 12) + 2 \times (12 \times 22) + 2 \times (7 \times 22)$

$= 168 + 528 + 308$

$= 1004$

Answer : 1004 cm^2 of carboard is needed.

b. Volume : $7 \times 12 \times 22 = 1848$

Answer : The box occupies 1848 cm^3.

Solve the problems. Show your work.

① How many cm^3 are there in 1 m^3?

Answer : _____

② How many cm^2 are there in 1 m^2?

Answer : _____

③ A cube-shaped water container has dimensions 12 cm by 12 cm by 12 cm. How many litres of water can it contain?

Answer : _____

④ The total surface area of a cube is 54cm^2. What are the dimensions of the cube?

Answer : _____

> **Read this first.**
>
> For Question 1,
> 1 m = 100 cm
> 1 m^3 =100 cm x 100 cm x 100 cm
>
> For Question 2,
> 1 m = 100 cm
> 1 m^2 =100 cm x 100 cm
>
> For Question 3,
> 1 L = 1000 mL; 1 mL =1 cm^3

ISBN: 978-1-897164-21-1

⑤ The Riverview Emporium is located at the intersection of River Street and Main Street. Sketch diagrams to show how the building would look from

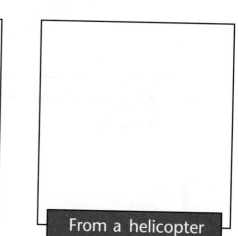
Riverview Emporium
River Street
Main Street

a. River Street.
b. Main Street.
c. a helicopter hovering over the building.

From River Street	From Main Street	From a helicopter

⑥ The figure shown is made of 8 identical cubes. The cube marked k is removed. What effect does this have on the total surface area of the figure? Explain.

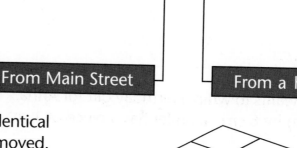

Answer : _____

⑦ The figure shown is made of 18 identical cubes, each having dimensions 2 cm by 2 cm by 2 cm.

a. If the cube marked x is removed, will the total surface area increase or decrease?

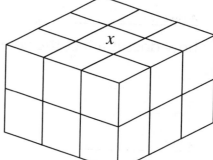

Answer : _____

b. How many cm^2 will the surface area increase or decrease?

Answer : _____

ISBN: 978-1-897164-21-1

⑧ Jill made a rectangular prism with tape, a pair of scissors and a piece of cardboard measuring 12 cm by 12 cm. Look how she cut and folded the cardboard to make the prism.

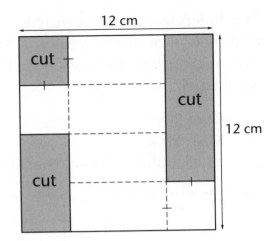

a. What are the dimensions of the prism?

Answer : _____

b. What is the total surface area of the prism?

Answer : _____

c. What is the volume of the prism?

Answer : _____

⑨ Jennifer wants to wrap a birthday gift for Anna. The dimensions of the box are 20 cm by 30 cm by 8 cm. Jennifer has a piece of 1 m² paper. Will she have enough paper?

Answer : _____

⑩ If you double each of the dimensions of a cube,

a. how does it affect the surface area?

Answer : _____

b. how does it affect the volume?

Answer : _____

⑪ The total surface area of a cube is 216 m². What is the volume of the cube?

Answer : _____

ISBN: 978-1-897164-21-1

⑫ Bill constructed a wooden composter measuring 1.5 m by 1.2 m by 1 m.

a. Determine the amount of wood required.

Answer : _____

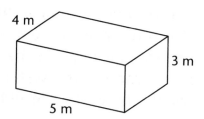

b. Determine the volume of compost it can hold.

Answer : _____

⑬ The dimensions of a room are 5 m by 4 m by 3 m. The area of the window on a wall is 3 m². If one 4 L can of paint can cover 36 m² and you want to paint the ceiling and the walls with 2 coats each, how many 4 L cans of paint do you need?

Answer : _____

⑭ How many dice each measuring 2 cm by 2 cm by 2 cm can be placed in a box of 10 cm by 10 cm by 10 cm?

Answer : _____

⑮ Ann has a collection of 50 hardcover books. 30 of the books have dimensions 16 cm by 23 cm by 3 cm and the other 20 books have dimensions 16 cm by 23 cm by 1.5 cm. What is the minimum volume of a container which will hold all these books?

Answer : _____

⑯ A box of laundry detergent (Box A) measuring 17 cm by 30 cm by 30 cm costs $11.99. Another box (Box B) measuring 15 cm by 25 cm by 25 cm costs $6.99. Which is a better buy? Explain.

Answer : _____

ISBN: 978-1-897164-21-1

⑰ A cereal box measures 31 cm by 20 cm by 7 cm. It is completely filled with cereal. How many servings of cereal does it contain if each serving has a volume of 175 mL?

Answer : _____

⑱ It takes a workman one hour to dig a 3 m by 3 m by 3 m hole. How long would 2 workmen take to dig a 6 m by 6 m by 6 m hole at the same rate?

Answer : _____

⑲ The volume of the pyramid is $\frac{1}{3}$ of the volume of water in the rectangular tank. If the pyramid is submerged in the water, how high will the new water level be?

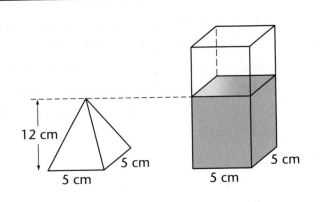

12 cm

5 cm

5 cm

5 cm

5 cm

Answer : _____

⑳ The concrete steps leading to John's house have the dimensions shown. Determine the volume of cement needed to build the steps.

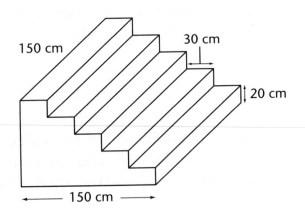

150 cm

30 cm

20 cm

150 cm

Answer : _____

㉑ A square piece of cardboard has an area of 25 cm². A shape of 1 cm² is cut from each corner. The sides are then folded to make an open box. What is the capacity of the box in mL?

cut   1 cm

1 cm

Answer : _____

ISBN: 978-1-897164-21-1

㉒ You are going to fill up the tank with tap water running at a rate of 20 L per minute. How long would it take?

1 m

1 m

0.5 m

1 m

1 m

3 m

Answer : _____

㉓ A box measures 6 cm by 4 cm by 6 cm. All the surfaces of the box have been painted. The box is then cut up into cubes measuring 1 cm³. How many of these small cubes will have just one face painted?

4 cm

6 cm

Answer : _____

CHALLENGE

Lennie tries to build different rectangular prisms with 12 1cm³ wooden cubes.

① Draw the different prisms he can build.

1

1

12

Answer : _____

② Calculate the surface area of each prism.

Answer : _____

③ Will the prisms have the same volume?

Answer : _____

Congruence and Similarity

EXAMPLE

Which of the triangles below are similar? Which are congruent?

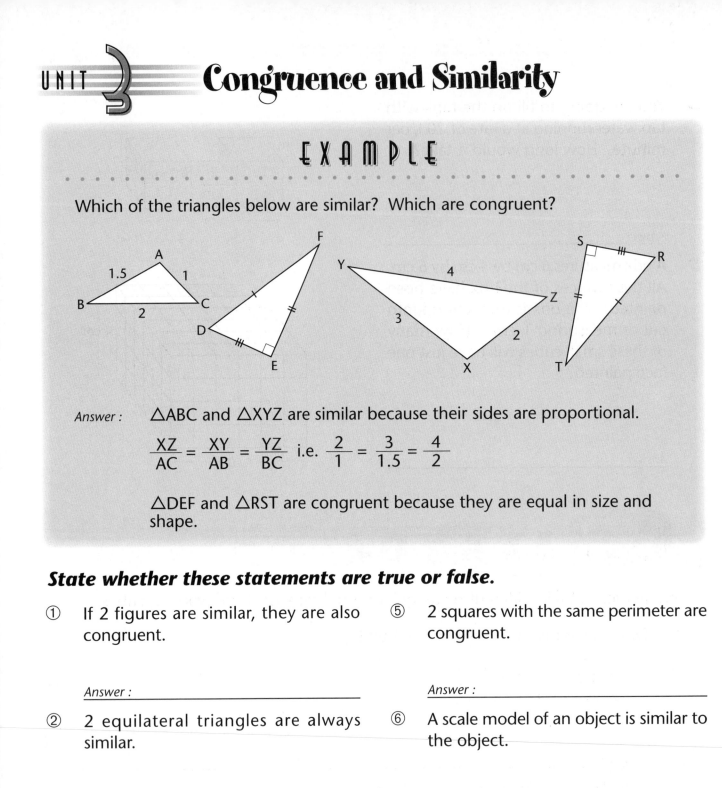

Answer :   △ABC and △XYZ are similar because their sides are proportional.

$$\frac{XZ}{AC} = \frac{XY}{AB} = \frac{YZ}{BC} \quad \text{i.e.} \quad \frac{2}{1} = \frac{3}{1.5} = \frac{4}{2}$$

△DEF and △RST are congruent because they are equal in size and shape.

State whether these statements are true or false.

① If 2 figures are similar, they are also congruent.

Answer : _____

② 2 equilateral triangles are always similar.

Answer : _____

③ 2 rectangles of the same perimeter are always similar.

Answer : _____

④ If 2 figures are congruent, they are also similar.

Answer : _____

⑤ 2 squares with the same perimeter are congruent.

Answer : _____

⑥ A scale model of an object is similar to the object.

Answer : _____

> • **2 figures are congruent if they are equal in size and shape.**   ◄ **Read this first.**
>
> • **2 figures are similar if they have the same shape but not necessarily the same size. Their corresponding angles are equal and their corresponding sides are proportional.**

ISBN: 978-1-897164-21-1

Solve the problems. Show your work.

⑦ An architect is drawing up the blueprint of Bill's house. The rectangular house has measurements 15 m by 18 m. If the scale of the blueprint is 1:300, determine the dimensions of the house on the blueprint.

Answer :
_____

⑧ Bill has a rectangular swimming pool that measures 3 m by 7 m. Bob has a swimming pool that measures 3.6 m by 8.4 m. Are the two swimming pools similar? Give reasons to support your answer.

Answer :
_____

⑨ There are 2 rectangular pictures on the wall of Bill's living room. One has dimensions 37 cm by 27 cm and the other, 26 cm by 21 cm. Are they similar? Explain.

Answer :
_____

⑩ The main span of New York's Brooklyn Bridge is 540 m long. On a blueprint it measures 6 cm. What is the scale factor of the blueprint?

Answer :
_____

⑪ Tom and Joe each construct a triangle with sides 4 cm , 5 cm and 6 cm. Will the triangles be congruent? Explain.

Answer :
_____

⑫ 2 rectangular stamps are similar to each other. The smaller stamp has 2 sides of 2 cm and 2 sides of 1.2 cm. The larger stamp has 2 sides of 3 cm. Determine the length of the other 2 sides. (Hint: there are 2 possible values.)

Answer :
_____

ISBN: 978-1-897164-21-1

⑬ Jonathan's jean pockets are similar. Determine the side length of the larger pocket marked X.

Answer : _____

⑭ Two similar trapezoidal tents are erected. Determine the length marked X and the size of angle A.

Answer : _____

⑮ The sphinx is 20 m high and 14 m wide. If a scale model is made with a height of 9.6 cm, what should the width be?

Answer : _____

⑯ Are the 2 cars shown similar to each other? Explain.

Answer : _____

⑰ 2 rectangular pictures are hung side by side as shown. Are the pictures similar to each other? Explain.

Answer : _____

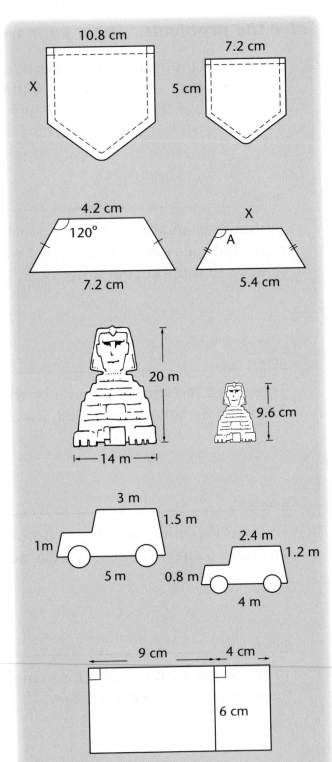

⑱ Ann and Betty each construct a triangle having angles 60°, 30° and 90°. Will the two triangles be congruent? Explain.

Answer : _____

⑲ An equilateral triangle has sides of 5 cm. Another equilateral triangle has sides of 4 cm. Are the triangles similar? Explain.

Answer : _____

ISBN: 978-1-897164-21-1

⑳ Divide the shape into 4 congruent figures each of which is similar to the original shape. What is the area of each shape?

Answer : _____

㉑ Describe as many similar triangles as you can in the map of Midtown Manhattan in New York City.

Answer : _____

㉒ The figure is called a tetrahedron.

a. What is the shape of each face?

Answer : _____

b. Are all the faces congruent?

Answer : _____

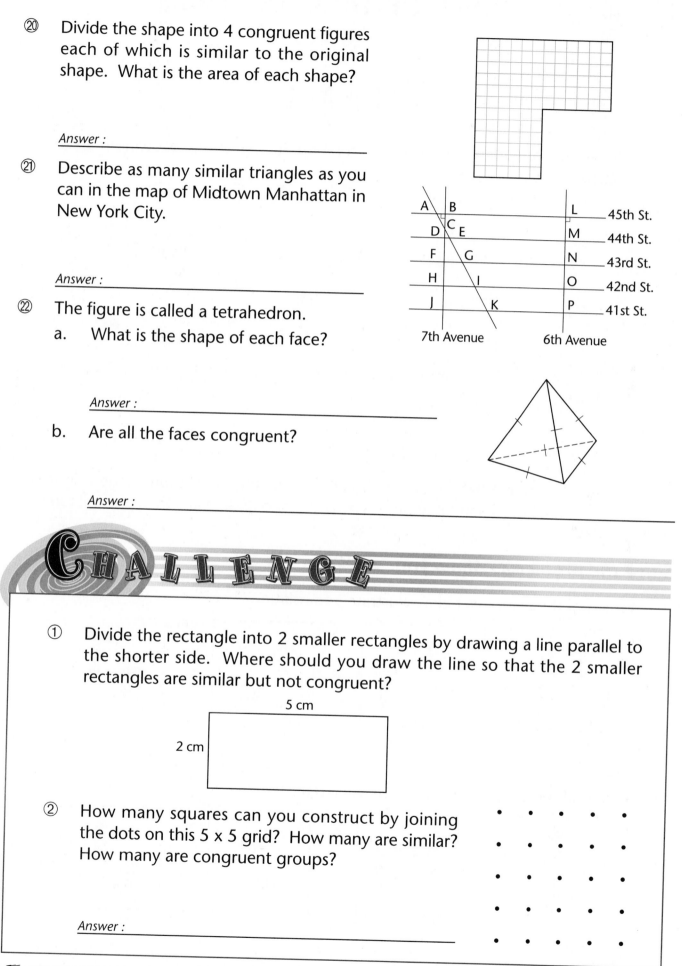

CHALLENGE

① Divide the rectangle into 2 smaller rectangles by drawing a line parallel to the shorter side. Where should you draw the line so that the 2 smaller rectangles are similar but not congruent?

5 cm

2 cm

② How many squares can you construct by joining the dots on this 5 x 5 grid? How many are similar? How many are congruent groups?

Answer : _____

ISBN: 978-1-897164-21-1

Transformations and Tiling

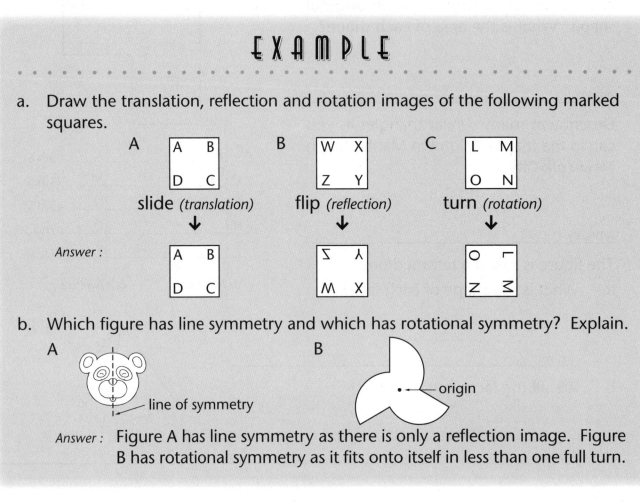

EXAMPLE

a. Draw the translation, reflection and rotation images of the following marked squares.

b. Which figure has line symmetry and which has rotational symmetry? Explain.

Answer : Figure A has line symmetry as there is only a reflection image. Figure B has rotational symmetry as it fits onto itself in less than one full turn.

Solve the problems. Show your work.

① Which of the diagrams below illustrate a translation?

A 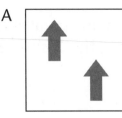   B 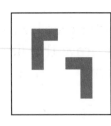   C 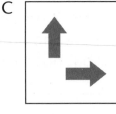   D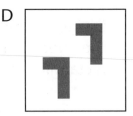

Answer : _____

② Which of the diagrams below illustrate a reflection?

A    B 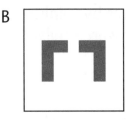   C    D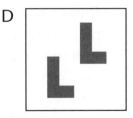

Answer : _____

ISBN: 978-1-897164-21-1

③ Which of the diagrams below illustrate a rotation?

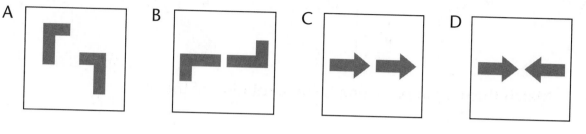

A   B   C   D

Answer : _____

④ Look how Bill moves a piece of glass with a letter B and a dot on it. Write the transformations.

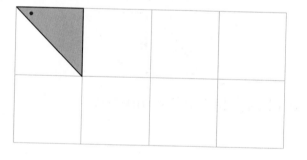

a. B. → B. → B. → B.

b. B. → ᗺ. → B. → ᗺ.

c. B. → ᗺ. → ᗺ. → B.

Answer :   a. _____ b. _____ c. _____

⑤ You are helping Bill create an imaginative design for his paved driveway. Use a transformation of your choice to create the pattern. Draw at least 6 additional triangles on the grid below. Describe the transformation used.

Answer : _____

⑥ The pattern on the wallpaper in Bill's bedroom is made up of congruent rectangles. Describe the transformation associated with the arrangement.

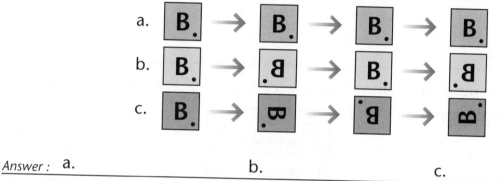

Answer : _____

ISBN: 978-1-897164-21-1

⑦ A traditional Canadian quilt contains the following 7 shapes.

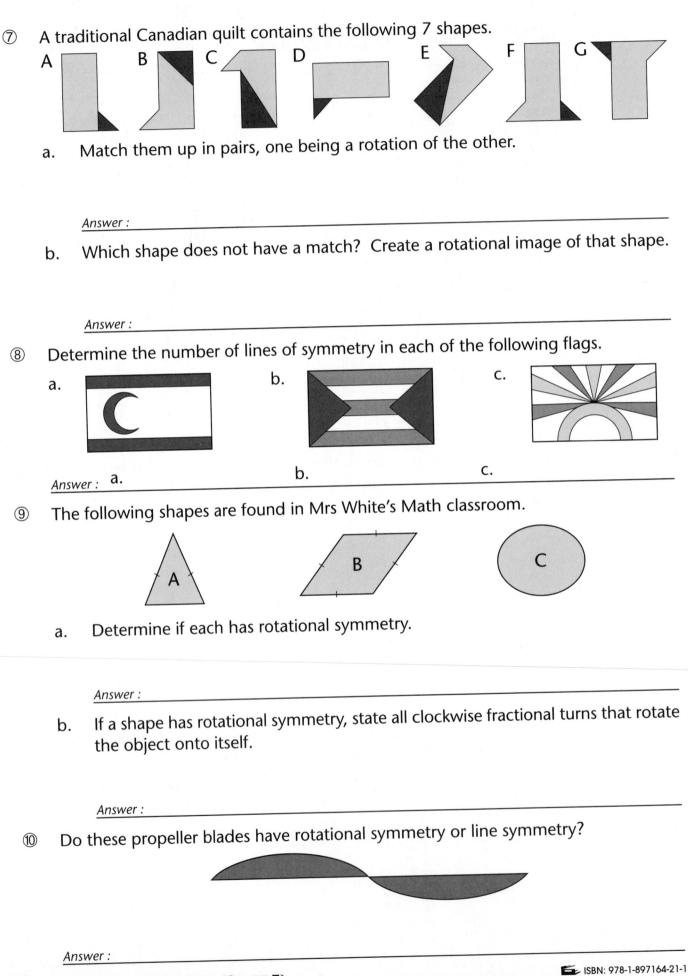

a. Match them up in pairs, one being a rotation of the other.

Answer : _____

b. Which shape does not have a match? Create a rotational image of that shape.

Answer : _____

⑧ Determine the number of lines of symmetry in each of the following flags.

a.

b.

c.

Answer : a. _____ b. _____ c. _____

⑨ The following shapes are found in Mrs White's Math classroom.

a. Determine if each has rotational symmetry.

Answer : _____

b. If a shape has rotational symmetry, state all clockwise fractional turns that rotate the object onto itself.

Answer : _____

⑩ Do these propeller blades have rotational symmetry or line symmetry?

Answer : _____

ISBN: 978-1-897164-21-1

⑪ The figure below has rotational symmetry. What is the order of the symmetry?

Read this first.

• A pattern which fits on itself in less than a complete rotation has rotational symmetry. If it fits on itself 3 times before it returns to its original position, it has rotational symmetry of order 4.

Answer : _____

⑫ Arrange the following figures in order from the highest to the lowest order of rotational symmetry.

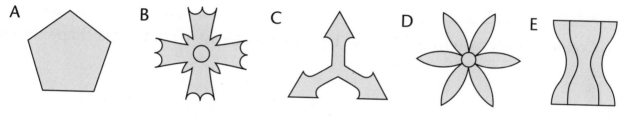

A B C D E

Answer : _____

⑬ Use a 90° rotation of the given design to create new figures in the squares shown.

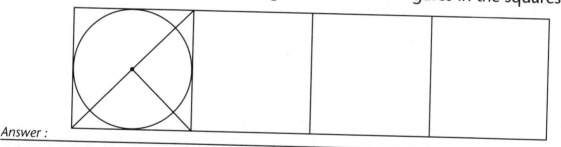

Answer : _____

⑭ Look at the following symbols.

a. Which symbols have a horizontal line of symmetry?

Answer : _____

b. Which symbols have a vertical line of symmetry?

Answer : _____

c. Which symbols have rotational symmetry?

Answer : _____

⑮ M.C. Escher made many pictures with different figures. Which of them can cover a flat surface without gaps or overlaps?

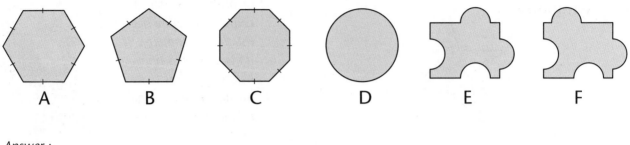

A   B   C   D   E   F

Answer : _____

⑯ Find at least 4 letters of the alphabet which can make a tiling pattern.

Answer : _____

⑰ Create a tiling pattern on the grid below with at least 10 of the tile shape shown.

Answer : _____

Use the geoboard or dot paper to answer each of the following questions.

⑱ Draw the images of triangle A on the geoboard

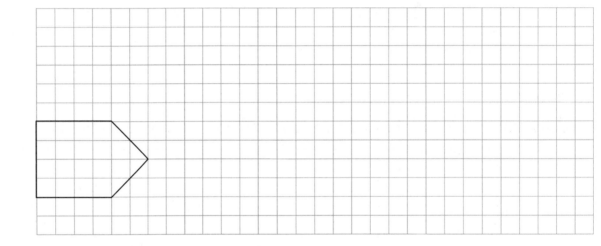

   a.   translated 5 right and 3 down
        (label the triangle B).

   b.   rotated 180° about the point *c*
        (label the triangle C).

   c.   reflected in the line *l* (label the
        triangle D).

   d.   reflected in the line *m* (label the
        triangle E).

ISBN: 978-1-897164-21-1

⑲ Peter has drawn 4 parallelograms on the dot paper as shown.

a. Compare the areas of the parallelograms.

Answer : _____

b. Describe the transformations which transfer K to L, K to M, and K to N.

Answer : _____

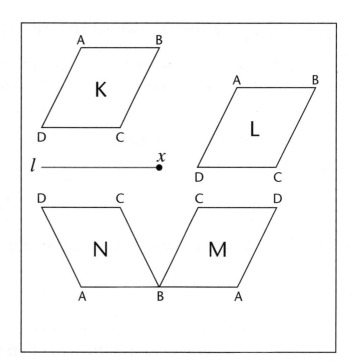

⑳ Ann has drawn a house on the dot paper. Draw the images of the house

a. rotated 180° about the point x (label the house B).

b. reflected in the line l (label the house C).

c. translated 6 right and 1 up (label the house D).

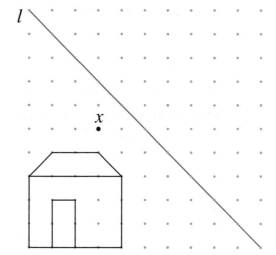

CHALLENGE

① Can a figure have line symmetry but not rotational symmetry? Explain with an example of a figure.

Answer : _____

② Draw a figure which has rotational symmetry of order 3 but does not have line symmetry.

Answer : _____

ISBN: 978-1-897164-21-1

Solve the problems. Show your work.

① Bill decides to build a scale model of the family cottage for his Technology project. The scale of the model is 1:100. The family cottage looks like a rectangular prism and its dimensions are 12 m by 12 m by 3 m.

a.   Determine the dimensions of the model.

Answer : _____

b.   How much space is there inside the model?

Answer : _____

c.   How much cardboard is needed to make the model? Do not consider the window and the door.

Answer : _____

d.   How many times is the space occupied by the cottage bigger than that occupied by the model?

Answer : _____

e.   How many times is the interior surface area of the cottage bigger than that of the model?

Answer : _____

f.   Which of the following nets can Bill use to construct his model? Circle the right answer.

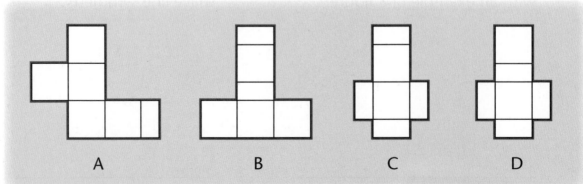

A   B   C   D

ISBN: 978-1-897164-21-1

② Peter has made a model of a barn with cardboard and 27 2-cm cubes. Look at his model and solve the problems.

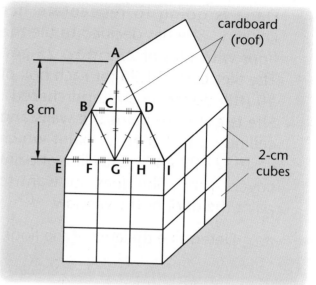

a. How many triangles are similar to △ABC?

Answer : _____

b. How many triangles are congruent to △ABD?

Answer : _____

c. Find the area of the parallelogram BDGE.

Answer : _____

d. Find the area of the trapezoid BDIE.

Answer : _____

e. If AI is 8.5 cm, what is the total surface area of the roof?

Answer : _____

f. What is the total surface area of the model?

Answer : _____

g. If Peter removes the part made with cardboard and paints the outside of the 6-cm cube, how many of the 2-cm cubes will have 2 painted faces?

Answer : _____

h. Peter submerged the 6-cm cube into the water in a rectangular container as shown. What was the original height of the water?

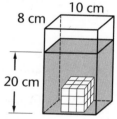

Answer : _____

i. Peter has 9 identical stickers. He has put one of them on a surface of the 6-cm cube. Help him put all the stickers onto the surface so that they are the reflection images of the sticker at the centre.

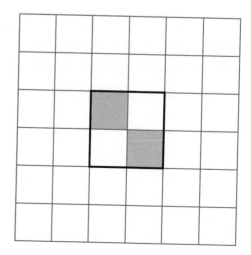

ISBN: 978-1-897164-21-1

③ Dana is going to redecorate her kitchen. She has decided to tile the floor with tiles of 25 cm by 25 cm. The tiles cost $21.95 for each box of 50 (full boxes must be purchased). She is going to paint the walls and ceiling with 2 coats of paint which costs $27.99 / 4 L. Each litre of paint covers 4 m². She does not want to paint the door or the window.

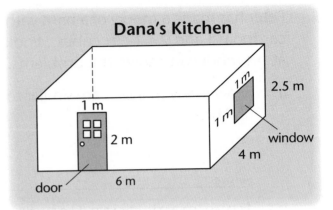

Dana's Kitchen

a. Determine the area of the floor.

Answer : _____

b. Determine the area of the walls and the ceiling (minus the areas of the door and the window).

Answer : _____

c. Determine the number of tiles needed to cover the floor.

Answer : _____

d. Determine the cost of tiling the floor.

Answer : _____

e. Determine the number of tiles left over.

Answer : _____

f. Determine the amount of paint needed to cover the walls and ceiling with 2 coats of paint.

Answer : _____

g. Determine the cost of the paint.

Answer : _____

h. Determine the amount of paint left over.

Answer : _____

④ How would you cut a cube three times to get 8 identical solid shapes which have 40 faces all together. Explain with the help of a diagram.

Answer : _____

ISBN: 978-1-897164-21-1

⑤ David drew 2 similar triangles as shown. Determine the perimeter of △XYZ.

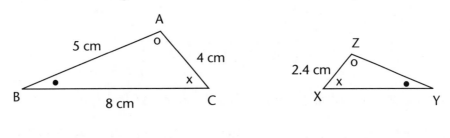

Answer : _____

⑥ You are going to tile the floor of your kitchen. Which of the following tiles would you not consider? Explain.

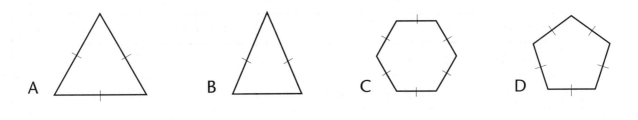

Answer : _____

⑦ Mrs Chan has many interlocking bricks which are regular octagons. She cannot make a tiling pattern with the bricks when she uses them to pave her driveway. What should she do? Draw the pattern.

Answer : _____

⑧ For each of the following diagrams, draw all the lines of symmetry. Then determine the order of rotational symmetry.

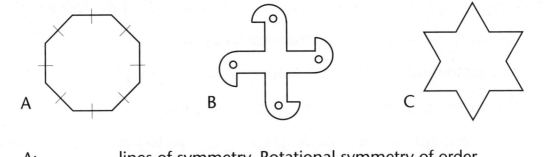

Answer : A: _____ lines of symmetry. Rotational symmetry of order _____ .

B: _____ lines of symmetry. Rotational symmetry of order _____ .

C: _____ lines of symmetry. Rotational symmetry of order _____ .

UNIT 5 Data Management

EXAMPLE

28 Grade 7 students recorded their weights in kg as follows:

$$42, 35, 37, 41, 45, 39, 40, 49, 48, 51, 40, 38, 36, 39,$$
$$45, 48, 40, 40, 50, 41, 42, 40, 39, 44, 47, 48, 50, 40$$

Make a tally chart to organize the data and graph the weights using a circle graph.

a. **Tally chart / Frequency table**

Weight (kg)	Tally	Frequency (No. of students)
30 – 39	//// //	7
40 – 49	//// //// //// ///	18
50 – 59	///	3

b.

Weight (kg)	Fraction	Angle
30 – 39	$\dfrac{7}{28} = \dfrac{1}{4}$	$\dfrac{1}{4} \times 360° = 90°$
40 – 49	$\dfrac{18}{28} = \dfrac{9}{14}$	$\dfrac{9}{14} \times 360° = 231°$
50 – 59	$\dfrac{3}{28}$	$\dfrac{3}{28} \times 360° = 39°$

Weights of Students

Circle the correct answer in each problem.

① What type of graph would you draw to represent each of the following?

a. Toronto's maximum temperature each day for a week.

    A. pictograph   B. line graph   C. bar graph   D. circle graph

b. The scores of all football teams in the World Cup.

    A. line graph   B. circle graph   C. pictograph   D. bar graph

c. The percentage of your weekly expenditure on different activities.

    A. circle graph   B. pictograph   C. bar graph   D. line graph

ISBN: 978-1-897164-21-1

② The circle graphs below show the fraction of Canada's population under 19 years old in the years 1900 and 1990. Which conclusion can you draw from the graphs?

A. The total population in Canada remained unchanged from 1900 to 1990.

B. The number of people under 19 years old decreased from 1900 to 1990.

C. The percentage of population under 19 years old decreased from 1900 to 1990.

D. There are more males than females under 19 years old.

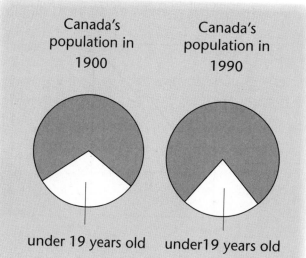

Canada's population in 1900

Canada's population in 1990

under 19 years old   under19 years old

Use the graphs to answer the questions.

③ The graph shows the percentage of female workers in a factory from 1990 to 2000.

a. When was there the greatest increase in the percentage of female workers?

Answer : _____

b. When was there a decrease in the percentage of female workers?

Answer : _____

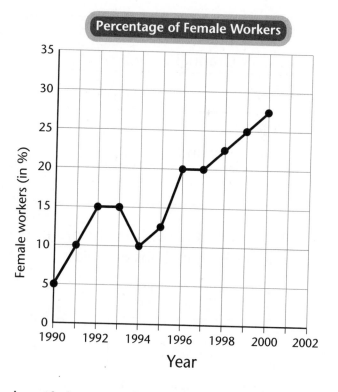

Percentage of Female Workers

c. Could a circle graph be used to represent the given data? Explain.

d. If the current trend continues, what percentage of female workers would be expected in the year 2002?

Answer : _____

Answer : _____

ISBN: 978-1-897164-21-1

④ Explain what is wrong with each of the following graphs and what message each graph is meant to convey.

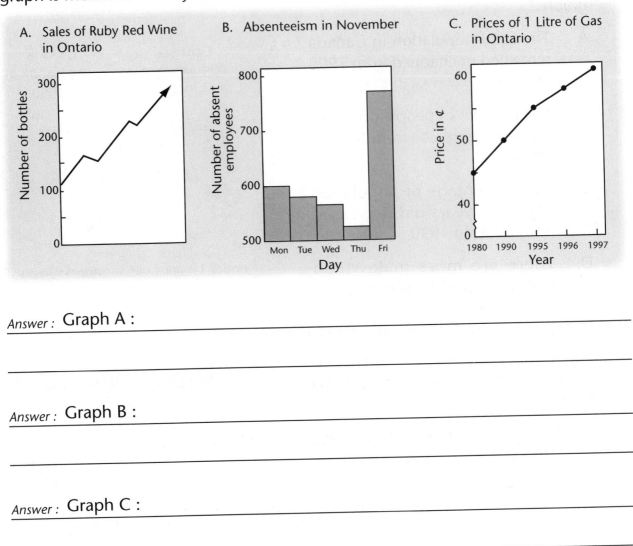

A. Sales of Ruby Red Wine in Ontario

B. Absenteeism in November

C. Prices of 1 Litre of Gas in Ontario

Answer : Graph A :

_____

_____

Answer : Graph B :

_____

_____

Answer : Graph C :

_____

_____

⑤ The graphs below show the profits earned by a certain drug company between 1990 and 1999 in millions of dollars.

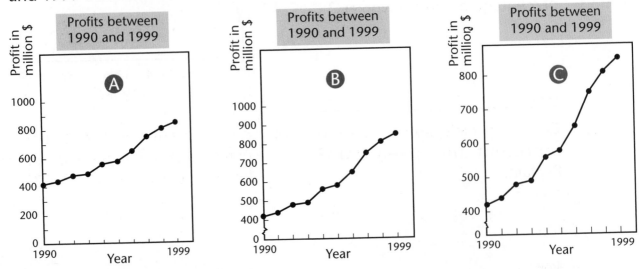

ISBN: 978-1-897164-21-1

a. Which of these graphs would you use if you were

   i. the President of the company at a meeting of shareholders? Explain.

   Answer : _____

   ii. a politician talking about excessive drug company profits? Explain.

   Answer : _____

   iii. the President of the company explaining to the press that drug company profits are not excessive?

   Answer : _____

b. Predict what the company profit might be in the year 2000.

   Answer : _____

c. Use the data below to draw a bar graph showing the profits made by the company between 1990 and 1999.

Year	Profit (in million $)
1990	420
1991	450
1992	490
1993	500
1994	530
1995	570
1996	620
1997	730
1998	800
1999	850

Profits between 1990 and 1999

⑥ 60 students were asked which form of exercise they preferred. The results are given in the chart below.

a. Complete the chart.

Activity	Number of students	Angle in a circle
Cycling	10	
Running	15	
Swimming	25	
Aerobics	10	

b. Draw a circle graph to illustrate the results.

c. What other type of graph could be used in this case?

Answer : _____

d. What type of graph would not be suitable?

Answer : _____

⑦ The following scores were recorded for a recent Math test.

58, 68, 72, 95, 37, 50, 75, 66, 83, 89, 48, 52, 64, 67, 75, 79, 85, 91

a. Organize the data into a stem-and-leaf plot.

b. Organize the data into a frequency table.

ISBN: 978-1-897164-21-1

c.   Graph the data using a bar graph.

d.   Graph the data using a circle graph.

e.   What percentage of students passed the test if the passing mark was 50?

Answer : _____

The graph shows the variation in the price of 1 oz of gold in US dollars in 1999.

①   When did gold reach its highest value and lowest value?

Answer : _____

②   What was the percentage change in the value of gold between January 1 and December 31, 1999?

Answer : _____

③   What was the percentage change from the lowest to the highest value?

Answer : _____

EXAMPLE

15 students recorded their heights in centimetres as follows:

150, 147, 155, 154, 160, 158, 148, 156, 155, 162, 170, 165, 163, 145, 153

Make a stem-and-leaf plot and use it to calculate the 3 measures of central tendency — mean (average), median (middle value) and mode (most common value).

Solutions :

tens	ones
14	5, 7, 8
15	0, 3, 4, 5, 5, 6, 8
16	0, 2, 3, 5
17	0

mean : $\dfrac{\text{sum of values}}{\text{number of values}}$

$= \dfrac{2341}{15}$

$= 156$

median : middle value
= 8th value (7 values below and 7 above)
= 155

mode : most common value
= 155 (occurs twice)

Solve the problems. Show your work.

① For each of the following statements, create a set of data with at least 3 values which will make the statement true.

a. The mean is smaller than the mode.

Answer : _____

b. The mode is smaller than the mean.

Answer : _____

c. The median is smaller than the mean.

d. The mean is smaller than the median.

Answer : _____

e. The mode and the median are the same.

Answer : _____

f. The mean of the values is 7.

Answer : _____

② The Toronto Hockey Team had the following number of shots on goal last month:

15, 20, 20, 15, 18, 23, 18, 24, 18

- If there is an even number of values, the <u>median</u> is the average of the 2 middle values.
- There may be more than 1 <u>mode</u> or none at all.

Read this first.

a. Determine the mean.

Answer : _____

b. Arrange the numbers in order and determine the median and the mode.

Answer : _____

c. If one additional game was played, with 29 shots on goal, would this affect the mean, median and/or mode? Explain.

Answer : The mean _____

_____ The median _____

_____ The mode _____

③ The stem-and-leaf diagram below shows the number of hours of TV viewing per week by a group of 13-year-olds in Toronto.

tens	ones	tens	ones
0	7, 9	3	5, 5, 5, 8, 8
1	1, 2, 4, 4, 8, 8, 9	4	2, 2
2	1, 3, 4, 6, 6, 8, 8		

a. Use the diagram to determine the mean of the number of hours of TV viewing per week. (correct to 2 decimal places)

Answer : _____

b. Determine the median and the mode.

Answer : _____

ISBN: 978-1-897164-21-1

④ The table below shows the annual salaries of the employees of ABC Construction Company.

Annual salary per person	Number of employees
$30 000.00	12
$40 000.00	2
$45 000.00	3
$60 000.00	2
$100 000.00	1 (President)

a. Calculate the mean, median and mode of the annual salaries.

Answer : _____

b. In an advertisement to attract new employees for the company, would you use the mean, median or mode? Why?

Answer : _____

c. If the President got a pay increase, would this affect the mean, median or mode?

Answer : _____

d. Which measure of central tendency best represents the given salaries? Why?

Answer : _____

⑤ The regular price of an ice cream cone is $1.25 but on Mondays it sells for $0.95. Karen and her friends bought 5 ice cream cones on Saturday, 3 on Sunday and 4 on Monday. Calculate the average price they paid.

Answer : _____

⑥ Dan and his classmates are talking about their weekly allowances for lunch. Dan has $25.00, 2 of his classmates have $32.00 each, 3 have $24.00 each, and 4 have $28.00 each. Jane brings sandwiches to school for lunch and has no allowance. Determine the mean of the allowances of the children. (correct to 2 decimal places)

Answer : _____

ISBN: 978-1-897164-21-1

⑦ Determine the mean, median and range of the following set of test results.

68, 70, 72, 78, 78

• **Range :** the difference between the highest and the lowest data values.

Read this first.

Answer : _____

⑧ The line graph shows the winning heights in Women's High Jump at the Olympic Games between 1968 and 1996. Use the graph to answer the following questions.

a. What trend do you notice in the winning heights?

Answer : _____

b. Calculate the approximate average winning height over the interval shown on the graph.

Answer : _____

c. Predict what the winning height might be in 2000.

Answer : _____

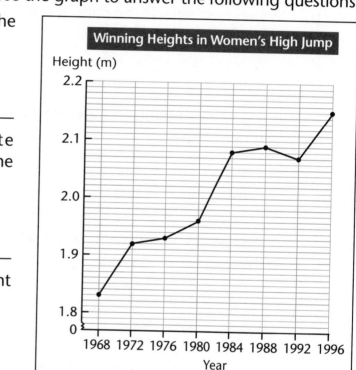

Winning Heights in Women's High Jump

Height (m)

2.2
2.1
2.0
1.9
1.8
0

1968 1972 1976 1980 1984 1988 1992 1996
Year

d. What is the range of these data values?

Answer : _____

⑨ A class of 25 students averaged 70% on a test. Another class of 30 students averaged 60%. Calculate the average percentage of all the 55 students.

Answer : _____

⑩ The average of 3 numbers is 20. One of the numbers is 15. Calculate the sum of the other 2 numbers.

Answer : _____

⑪ The table shows the number of oil tanker spills worldwide between 1982 and 1990.

Year	1982	1983	1984	1985	1986	1987	1988	1989	1990
Number of spills	9	17	15	9	8	12	13	31	8

a. Determine the mean, median and mode(s).

Answer : The mean   ; the median   ; the mode(s)   .

b. Which best reflects the data, the mode(s) or mean? Explain your answer.

Answer : _____

c. If you were a member of an environmental organization, which measure would you use? Why?

Answer : _____

d. If you were the president of an oil company, which measure would you use? Why?

Answer : _____

⑫ The annual sales of Evesview Company are shown on the bar graph.

a. Use the graph to determine the average (mean) sales of the company between 1991 and 1995.

Answer : _____

b. The graph shows a rapid increase in annual sales. How is this impression created?

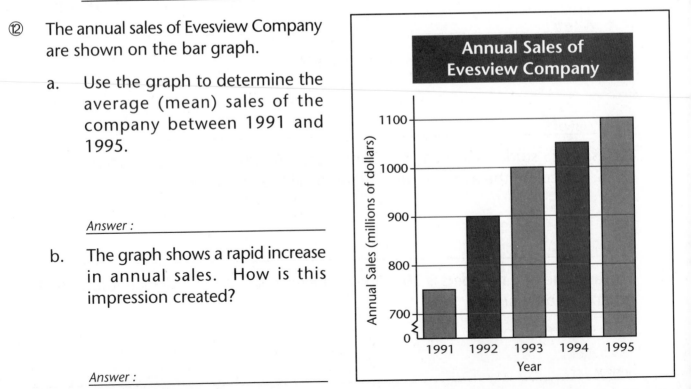

Annual Sales of Evesview Company

Answer : _____

ISBN: 978-1-897164-21-1

⑬ The table below shows the population density of some East Asian countries in 1992. Use the table to answer the questions.

a. Calculate the population density of each country to complete the table.

Country	Area (thousands of sq. km)	Population (millions)	Population Density ($\frac{Population}{Area}$)
Indonesia	1882	197	
Japan	374	125	
Philippines	297	67	
North Korea	120	22	
South Korea	97	44	

b. Which of these countries is the most densely populated?

Answer : _____

c. Determine the mean population size of these countries (in millions).

Answer : _____

d. Determine the median population size of these countries (in millions).

Answer : _____

e. Which type of graph could be used to represent these statistics? (Do not draw the graph.)

Answer : _____

CHALLENGE

7 students' test scores are recorded with a mean of 55%, a median of 60% and a mode of 70%. If the lowest score is 34% and the highest score is 71%, determine the 5 possible groups of test scores.

Answer : A. _____   D. _____

B. _____   E. _____

C. _____

ISBN: 978-1-897164-21-1

EXAMPLE

Jimmie wants to buy a new bike. He can choose a mountain bike or an all-terrain bike. The colours available are black, silver, white and gold. If all choices are equally likely,

a. in how many ways can Jimmie choose a new bike?

b. what is the probability that his bike will be black?

c. what is the probability that it will be an all-terrain bike?

d. what is the probability that it will be a black all-terrain bike?

Solutions :

Tree Diagram

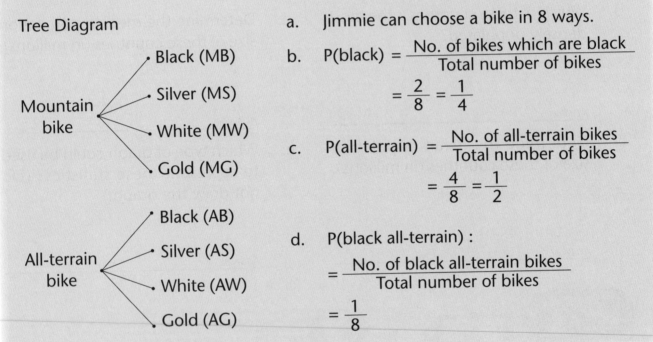

a. Jimmie can choose a bike in 8 ways.

b. P(black) = $\dfrac{\text{No. of bikes which are black}}{\text{Total number of bikes}}$

$= \dfrac{2}{8} = \dfrac{1}{4}$

c. P(all-terrain) = $\dfrac{\text{No. of all-terrain bikes}}{\text{Total number of bikes}}$

$= \dfrac{4}{8} = \dfrac{1}{2}$

d. P(black all-terrain) :

$= \dfrac{\text{No. of black all-terrain bikes}}{\text{Total number of bikes}}$

$= \dfrac{1}{8}$

Solve the problems. Show your work.

① Ann flipped a coin 50 times and heads turned up 24 times.

a. What fraction of the flips turned up heads?

Answer : _____

b. Is this what you would expect?

Answer : _____

ISBN: 978-1-897164-21-1

c. If the coin is fair and Ann flips the coin 500 times, how many times should she expect it to come up heads?

Answer : _____

d. What fraction of the flips should come up heads if she flips the coin many times?

Answer : _____

e. If the first 5 flips are all heads, what is the probability that the next flip will be heads?

Answer : _____

f. Is it possible to flip a coin 20 times and have it turn up heads every time? Is this likely to happen?

Answer : _____

② Carol and Debbie play a coin-tossing game. They take turns tossing 2 coins. If both coins match, Carol gets a point. If the coins don't match, Debbie gets a point. The first player to get 10 points wins. Who is likely to win? Illustrate your answer with a tree diagram in the box.

Answer : _____

③ Eric and Frank also play a coin-tossing game but they use 3 coins. If 3 coins match, Eric gets a point. If only 2 coins match, Frank gets a point. The first to score 10 points wins. Who is likely to win? Illustrate your answer with a tree diagram in the box.

Answer : _____

④ George spun each of the spinners below 30 times.

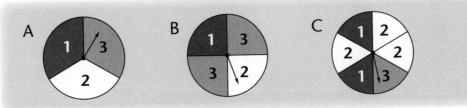

a. What is the probability that the result will be a 3 in each case?

Answer : _____

b. How many times should he expect each spinner to land on a 3?

Answer : _____

c. Does the probability of landing on a 3 depends on the number of times he spun each spinner?

Answer : _____

d. Which of the spinners would give a fair game of chance?

Answer : _____

⑤ In a TV guessing game, a bucket is filled with an unknown number of coloured balls. Members of the audience are asked to guess and write down, one after the other, the colour of the ball that would be drawn. If the guess is correct, they win a prize.

a. Would it be better to write first or last? Explain.

Answer : _____

b. If the bucket contains 10 red, 20 white and 30 black balls, what is the probability that a white ball will be drawn?

Answer : _____

c. What is the most likely outcome if the ball is drawn randomly?

Answer : _____

ISBN: 978-1-897164-21-1

⑥ When I toss a fair die,

a. what is the probability that the die will show a 4?

b. what is the probability that the die will show a number greater than 4?

Answer : _____

Answer : _____

⑦ Mrs Smith has made 2 spinners to help determine what she will make for lunch.

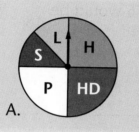

A.   B.

S – spaghetti
P – pizza
H – hamburger
HD – hot dog
L – lasagna

a. What is the probability that there will be hamburgers for lunch if only spinner B is spun?

Answer : _____

b. If you like spaghetti, which spinner would you like Mrs Smith to use?

Answer : _____

c. Frankie uses spinner A to predict the outcome and gets points if his prediction is correct. He can get 2 points for guessing that the outcome is hamburger. How many points should he get for guessing lasagna? Explain.

Answer : _____

⑧ When 2 dice are rolled and the numbers on the upper faces are added, there are 12 different possible sums.

a. Complete the sum chart below to show the possible outcomes.

⊕	1	2	3	4	5	6
1						
2						
3						
4						
5						
6						

b. Are all the different sums equally likely?

Answer : _____

c. What is the probability that the sum on the 2 dice is 4?

Answer : _____

d. Which sum is most likely?

Answer : _____

ISBN: 978-1-897164-21-1

⑨　In Monopoly, you get out of jail if you roll a double with 2 dice. What is the probability that you roll a double?

Answer : _____

⑩　Trish has 5 different polyhedral dice. They have 4, 6, 8, 12 and 20 faces respectively. She rolls any 2 dice together and records the sums of the faces that come up.

a.　She finds that the probability of a sum of 6 would be $\frac{5}{48}$. Which 2 dice is she using? Explain.

Answer : _____

b.　Using the 2 dice you found in a, determine the probability of getting a sum of 13.

Answer : _____

⑪　On your next family vacation to Quebec, you plan to visit Quebec city, Montreal and Hull.

a.　In how many different orders can you visit the 3 cities?

Answer : _____

b.　In practice, would all orders be equally likely? Explain.

Answer : _____

⑫　Sylvia rolls 3 dice.

a.　What is the probability that all 3 dice show a 6?

Answer : _____

b.　What is the probability that all 3 dice show 4 or 5?

Answer : _____

ISBN: 978-1-897164-21-1

c.   What is the probability that all dice match?

Answer :

⑬   A betting game consists of rolling a die. You only win if you roll a 6 and the prize is $2. How much money do you expect to win by rolling the die 60 times?

Answer :

⑭   5 cards marked 1, 2, 3, 4 and 5 are in a bag. If 2 cards are taken from the bag, what is the probability that

a.   both of them are even?

Answer :

b.   only one of them is even?

Answer :

⑮   A scratch and match card is made with 4 scratch-off spots. 2 of the spots have the same letter and the other 2 have different letters. To win, you must scratch only 2 spots and they must have the same letter. What is the probability that you will win?

Answer :

CHALLENGE

A CD club offers 4 monthly choices. You can choose any 2 CDs at a special discount price. In how many different ways can you choose 2 CDs?

Answer :

ISBN: 978-1-897164-21-1

Circle the correct answer in each problem.

① The perimeter of the larger square is 20 cm and the perimeter of the smaller square is 16 cm. What is the area of the region between the 2 squares?

A. 1 cm^2 B. 2 cm^2

C. 4 cm^2 D. 9 cm^2

② In a club of 24 members, each serves on three 4-person committees. How many committees are there?

A. 24   B. 32

C. 18   D. 12

③ The average of 3 numbers is between 7 and 10. The sum of the numbers could be any of the following EXCEPT

A. 22.   B. 20.

C. 28.   D. 26.

④ If each side of a square is increased by 50%, then the area of the square is increased by

A. 50%.   B. 100%.

C. 125%.   D. 225%.

⑤ You are joining a book club. You get 2 free choices from 4 books. How many different choices do you have?

A. 24   B. 8

C. 6   D. 12

⑥ The average (mean) of the first 10 positive whole numbers is

A. 5.   B. 5.5.

C. 6.   D. 10.

⑦ If 14 pigs equal 35 hogs, then 50 hogs equal

A. 125 pigs.   B. 70 pigs.

C. 20 pigs.   D. 7 pigs.

⑧ Jack takes 3 pairs of jeans, 4 T-shirts and 2 pairs of shoes for his trip. How many different outfits could he wear?

A. 9   B. 24

C. 12   D. 22

⑨ Ann, Bill, Carol and Dave serve on a school committee. In how many different ways could a President and Vice-President be chosen from the committee members?

A. 12   B. 6

C. 7   D. 8

⑩ How many squares are there in the following figure?

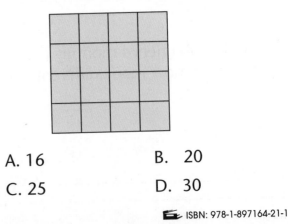

A. 16   B. 20

C. 25   D. 30

ISBN: 978-1-897164-21-1

The table below shows Julie's Math test scores so far. Use the table to solve the problems. Show your work.

Test	1st	2nd	3rd	4th	5th	6th
Score (%)	85	82	91	72	75	85

⑪ Julie's teacher lets Julie choose the mean, median or mode of the scores to be recorded on her interim report card. Which one should Julie choose? Explain.

Answer :

⑫ There are 3 more tests to write this semester. How can Julie get a median of 85 for her 9 tests?

Answer :

⑬ How can she get a mean of 80 for her 9 tests?

Answer :

⑭ What type of graph could be used to represent Julie's test scores? Explain.

Answer :

⑮ If Julie's teacher picks a percentage at random from the 6 tests written so far, what is the probability that the grade will be a B (70 - 79%)?

Answer :

⑯ Mary is Julie's classmate. She got 83 on the 2nd Math test. What is her performance on the other 5 tests if she has the same median as Julie for her first 6 tests?

Answer :

⑰ What is Mary's performance if she has the same mode as Julie on her first 6 Math tests?

Answer :

ISBN: 978-1-897164-21-1

⑱ Read the circle graph showing the time spent per day by a 12-year-old child on various activities.

a. According to the graph, how much time does a typical 12-year-old child spend on homework per day?

Answer : _____

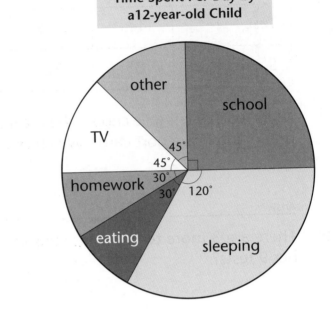

Time Spent Per Day by a 12-year-old Child

(circle graph labeled: other, school, TV, 45°, 45°, homework, 30°, 30°, 120°, eating, sleeping)

b. What percentage of a day is spent on sleeping? (correct to the nearest 0.1%)

Answer : _____

c. Why is a circle graph the most suitable for this application?

Answer : _____

d. How might the circle graph differ for an 18-year-old teenager?

Answer : _____

Solve the problems. *Show your work.*

⑲ Briana is the marketing manager of a chocolate company. In order to promote a new brand of chocolate bar, she is planning a prize system whereby every 50th chocolate bar will contain a 50¢ off coupon and every 200th chocolate bar will contain a $2.00 discount coupon. No chocolate bars can contain 2 prizes.

a. Explain how she should distribute the prizes fairly in 500 boxes of chocolate bars, each box containing 20 bars.

Answer : _____

b. If Jim buys 10 chocolate bars, what is the probability that he will win something?

Answer : _____

ISBN: 978-1-897164-21-1

c.   If all the chocolate bars are sold and 80% of the prizes are claimed, how much will this promotion cost the company?

Answer : _____

⑳   The Richmond School Athletic Council held a casino night to raise money for school sports.

a.   At the first table, the game consisted of spinning 3 spinners. Each spinner had a red, black, white and yellow sector. Jane tested each of the spinners by spinning 100 times and recorded the results on a graph. Use the graphs to help you decide what each of the spinners looked like. Explain your answers.

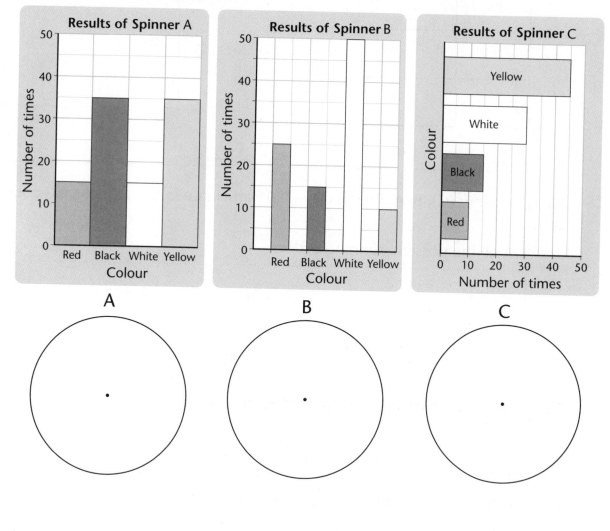

Answer : _____

b.   At another table, the game consisted of tossing 2 coins. You only win if you can toss 2 heads. Brian tossed the pair of coins 20 times. How many times should he expect to win?

Answer : _____

ISBN: 978-1-897164-21-1

㉑ 25 Grade 7 students recorded their ages in months as follows:

150　140　145　153　141　144　146　139　156　158　150　144　142
140　148　149　159　150　151　150　138　140　135　139　138

a. Make a stem-and-leaf plot and use it to determine the median age of the Grade 7 students.

b. Draw a suitable graph to show the ages of the Grade 7 students.

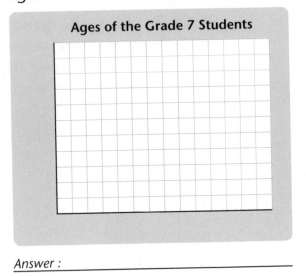

Ages of the Grade 7 Students

Answer : _____

Answer : _____

㉒ Jim (J), Charles (C), Ben (B) and Dave (D) participate in the school cross country relay team.

a. List all the possible running orders.

b. How many different running orders are there?

Answer : _____

c. What is the probability that Jim will run first?

Answer : _____

d. If Charles is the fastest runner and should therefore run the last leg, how many different running orders are there?

Answer : _____

ISBN: 978-1-897164-21-1

Handy Reference

1. Squares and Square Roots

- $\sqrt{a \times b} = \sqrt{a} \times \sqrt{b}$, e.g. $\sqrt{9 \times 4} = \sqrt{9} \times \sqrt{4} = 3 \times 2 = 6$
- $\sqrt{\dfrac{a}{b}} = \dfrac{\sqrt{a}}{\sqrt{b}}$, e.g. $\sqrt{\dfrac{9}{4}} = \dfrac{\sqrt{9}}{\sqrt{4}} = \dfrac{3}{2} = 1\dfrac{1}{2}$

2. Exponents

- $a^{-b} = \dfrac{1}{a^b}$, e.g. $3^{-2} = \dfrac{1}{3^2} = \dfrac{1}{9}$
- $a^m \times a^n = a^{m+n}$, e.g. $2^2 \times 2^4 = 2^{2+4} = 2^6$
- $a^m \div a^n = a^{m-n}$, e.g. $2^5 \div 2^3 = 2^{5-3} = 2^2$

3. Percents

- Discount   = regular price x discount rate

  Sale price = regular price − discount
- Tax   = price x tax rate

  Total cost = price + tax

Discount: $80 x 20% = $16

Sale price : $80 − $16 = $64

* 20% off   Tax : $64 x 15% = $9.60

* 15% tax   Total cost: $64 + $9.60 = $73.60

- Commission = total sales x rate of commission

  e.g. Jasmine earns 4% commission on each item she sells. If she sells $120 worth of items today, how much is her commission?

      Commission = $120 x 4%

                 = $4.80

  ∴ Jasmine's commission is $4.80.
- Interest

  $I = PRT$, I = interest

               P = principal

               R = interest rate

               T = time in year

  e.g. Alice deposits $1000 at an interest rate of 3% per year for 2 years. How much interest will she get after 2 years?

     I = $1000 x 3% x 2

       = $60

  ∴ Alice will receive $60 after 2 years.

4. Speed

- $S = \dfrac{d}{t}$, S = speed

            d = distance

            t = time

  e.g. Winnie walks 10 m in 20 s. What is her speed?

     $S = \dfrac{10 \text{ m}}{20 \text{ s}}$

      = 0.5 m/s

  ∴ Winnie's walking speed is 0.5 m/s.

ISBN: 978-1-897164-21-1

Handy Reference

5. Measurement Units Conversion

1 km = 1000 m	1 kg = 1000 g	1 L = 1000 mL
1 m = 100 cm	1 g = 1000 mg	1 mL = 1 cm^3
1 cm = 10 mm		1 cm^3 = 0.001 L
1 m^2 = 10 000 cm^2	e.g. A pill is about	e.g.
1 m^3 = 1 000 000 cm^3	50 mg.	2 L (2000 mL)

6. Perimeter, Area, Surface Area, and Volume

Perimeter = 4s
Area = s^2

Perimeter = 2(l + w)
Area = l x w

Area = $\frac{1}{2}$bh

Area = bh

Area = $\frac{1}{2}$(a + b)h

For any prism:
Volume = Base Area x Height

For any pyramid:
Volume = $\frac{1}{3}$ x Base Area x Height

For any prism and pyramid:
Surface Area = Combined area of all 2-D surfaces of the solid

Volume = $\pi r^2 h$
Surface Area = $2\pi rh + 2\pi r^2$

Volume = $\frac{1}{3}\pi r^2 h$
Surface Area = $\pi rs + \pi r^2$

7. Circle and Sphere

Diameter = radius x 2
Circumference = πd
= $2\pi r$

Area = πr^2

Volume = $\frac{4}{3}\pi r^3$
Surface Area = $4\pi r^2$

8. Pythagorean Theorem

$a^2 + b^2 = c^2$

e.g.

3 cm, 1 cm, k cm

$k^2 = 1^2 + 3^2$
$k^2 = 1 + 9$
$k^2 = 10$
$k = \sqrt{10}$
∴ The unknown side is $\sqrt{10}$ cm.

9. Slope

Slope = $\frac{rise}{run}$

e.g. (5,6) rise (1,0) run

Slope = $\frac{6 - 0}{5 - 1} = \frac{6}{4} = \frac{3}{2} = 1\frac{1}{2}$

ISBN: 978-1-897164-21-1

1 Number Theory

1. 4^5 ; 4 ; 5 ; 1024
2. 3^3 ; 3 ; 3 ; 27
3. 6^3 ; 6 ; 3 ; 216
4. 2 x 2 x 2 x 2 ; 2 ; 4 ; 16
5. 7 x 7 ; 7 ; 2 ; 49
6. 3 x 3 x 3 x 3 x 3 x 3 ; 3^6 ; 729
7. 8 x 8 x 8 x 8 ; 8^4 ; 4096
8. 10 000
9. 729
10. 1
11. 343
12. 16
13. 279
14. 32
15. 512
16. 1
17. <
18. <
19. >
20. =
21. <
22. <
23. 2^1 , 2^2 , 2^3 , 2^4
24. 1^5 , 3^2 , 2^7 , 5^7
25. 2 ; 9 ; 5 ; 6
26. 10^4 ; 10^3 ; 10^2 ; 10^1 ; 10^0
27. $7 \times 10^3 + 3 \times 10^2 + 6 \times 10^1 + 1 \times 10^0$
28. 946 912
29. 803 006
30. 800 ; 70 ; 5 ; $8 \times 10^2 + 7 \times 10^1 + 5 \times 10^0$
31. $7000 + 100 + 80 = 7 \times 10^3 + 1 \times 10^2 + 8 \times 10^1$
32. $1000 + 20 + 4 = 1 \times 10^3 + 2 \times 10^1 + 4 \times 10^0$
33. $50\ 000 + 9000 + 3 = 5 \times 10^4 + 9 \times 10^3 + 3 \times 10^0$

2 Squares and Square Roots

1. 5
2. 6
3. 9
4. 0
5. 1
6. 4
7. 144
8. 1225
9. 2401
10. 100
11. 256
12. 196
13. ≠
14. =
15. ≠
16. ≠
17. =
18. =

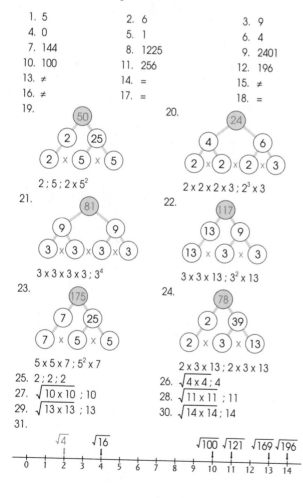

19. 2 ; 5 ; 2×5^2
20. $2 \times 2 \times 2 \times 3$; $2^3 \times 3$
21. $3 \times 3 \times 3 \times 3$; 3^4
22. $3 \times 3 \times 13$; $3^2 \times 13$
23. $5 \times 5 \times 7$; $5^2 \times 7$
24. $2 \times 3 \times 13$; $2 \times 3 \times 13$
25. 2 ; 2 ; 2
26. $\sqrt{4 \times 4}$; 4
27. $\sqrt{10 \times 10}$; 10
28. $\sqrt{11 \times 11}$; 11
29. $\sqrt{13 \times 13}$; 13
30. $\sqrt{14 \times 14}$; 14
31.

$\sqrt{4}$ at 2, $\sqrt{16}$ at 4, $\sqrt{100}$ at 10, $\sqrt{121}$ at 11, $\sqrt{169}$ at 13, $\sqrt{196}$ at 14 on number line 0 to 14.

32. 13
33. 4 ; 10
34. 13 ; 14
35. 10 ; 11

3 Multiples and Factors

1. Factors of 12: 1 , 2 , 3 , 4 , 6 , 12 ;
   Factors of 20: 1 , 2 , 4 , 5 , 10 , 20 ;
   Common factors: 1 , 2 , 4 ;
   G.C.F.: 4
2. Multiples of 6: 6, 12, 18, 24, 30, 36, 42, 48, 54, 60, 66, 72, 78, 84, 90, 96 ;
   Multiples of 15: 15 , 30 , 45 , 60 , 75 , 90 ;
   Common multiples: 30 , 60 , 90 ;
   L.C.M.: 30
3. 2 ; 3 ; 6
4. 5 ; 7 ; 35
5. 8
6. 5
7. 4 ; 4 ; 16
8. 3 ; 4 ; 12
9. 4 ; 3 ; 24
10. $3 \times 2 \times 5 \times 6$; 180
11. $6 \times 3 \times 8$; 144
12. $5 \times 2 \times 3 \times 4 \times 5$; 600
13. $5 \times 2 \times 3 \times 1 \times 1 \times 3$; 90
14. $4 \times 2 \times 3 \times 1 \times 1 \times 1$; 24

4 Integers

1. Circle the integers: -15 ; 7 ; -6 ; 120 ; 29
   Opposites: +15 ; -7 ; +6 ; -120 ; -29
2. +45 kg
3. -$200
4. -12° C
5. +140 m
6. +$35
7. -2 kg
8.-12. (Suggested answers)
8. -1
9. -6
10. 0
11. 0
12. -9
13. -6
14. +5
15. -38 , -14 , -4 , 6 , 19
16. -4 m , -3 m , 2 m , 15 m , 16 m
17. -15 , -3 , 0 , 11 , 23
18. -7° C , -5° C , 0° C , 9° C , 32° C
19. A
20. B
21. A
22. B
23. C
24. C
25. +9
26. -5
27. -9

ISBN: 978-1-897164-21-1

28. -10
29. -8
30. +4
31. +1
32. -4
33. +2
34. -7
35. +7
36. +1

37. (+7) + (+1) ; +8
38. (-7) + (-11) ; -18
39. (-13) + (+2) ; -11
40. (-9) + (-7) ; -16
41. (-5) ; -12
42. (-1) + (+6) – (-8) ; (+5) + (+8) ; +13
43. (+8) – (+12) + (-3) ; (-4) + (-3) ; -7
44. (-7) – (-8) – (-9) ; (+1) + (+9) ; +10
45. (-10) + (-6) – (+4) ; (-16) + (-4) ; -20
46. (+4) – (-8) + (-3) ; (+12) + (-3) ; +9

47. -20
48. +63
49. +42
50. -6
51. -36
52. -40
53. -36
54. +18
55. 0
56. -96
57. +5
58. -5
59. -3
60. +2
61. +3
62. -7
63. -4
64. -4
65. -1
66. -2
67. (-5) ; -13

68. (-10) + (-6) ÷ (+2) ; (-10) + (-3) ; -13
69. (-9) ÷ (-3) + (-9) ; (+3) + (-9) ; = -6
70. (-14) ÷ (-2) + (-6) ; (+7) + (-6) ; = +1
71. (-4) + (-2) x (-4) ; (-4) + (+8) ; = +4
72. (-3) x (-6) ÷ (-2) ; (+18) ÷ (-2) ; = -9
73. (-25) + (-9) x (-2) ; (-25) + (+18) ; = -7
74. (-2) – (-12) ÷ (-6) ; (-2) – (+2) ; = -4
75. (-8) ÷ (-4) x (-10) ; (+2) x (-10) ; = -20

5 Decimals

1. F
2. T
3. T
4. F
5. F
6. T
7. 0.54
8. 120
9. 207.5
10. 14 600
11. 300.5
12. 6.8
13. 0.2889
14. 116.4
15. 0.005867
16. 0.12
17. 0.00219
18. 0.548
19. 0.84
20. 0.2413
21. 19 010
22. 0.00315
23. 77.93
24. 91.94
25. 93.77
26. 101.02
27. 75.61
28. 610.28
29. 394.13
30. 2.854
31. 699.27
32. 76.28 ; 69.56
33. (21.42 – 16.9) + 7.65 ; 4.52 + 7.65 ; 12.17
34. (722.9 – 69) + 409.12 ; 653.9 + 409.12 ; 1063.02
35. (26.99 – 18.12) – 8.86; 8.87 – 8.86 ; 0.01
36. (56.45 + 47.58) – 20.98 ; 104.03 – 20.98 ; 83.05
37. (600 – 124.8) – 469.91 ; 475.2 – 469.91 ; 5.29
38. 5.46
39. 0.4772
40. 0.04266
41. 0.864
42. 0.8008
43. 1012 ; 2530 ; 0.03542
44. 27.208
45. 3.888
46. 333.12
47. 0.672
48. 1.0944
49. 89.18
50. $7.99 x 3 = $23.97 ; Mary pays $23.97.
51. $8.99 x 2 = $17.98 ; Judy pays $17.98.

52. $23.97 + $17.98 = $41.95 ; Mary's mother needs to pay $41.95.
53. $50 – $41.95 = $8.05 ; Her change is $8.05.
54. 1.8

$$6 \overline{)\, 10.8} \quad \begin{array}{r} 1.8 \\ \hline 6 \\ 48 \\ 48 \end{array}$$

55. 76.8 ÷ 8 ; 9.6

$$8 \overline{)\, 76.8} \quad \begin{array}{r} 9.6 \\ \hline 72 \\ 48 \\ 48 \end{array}$$

56. 6250 ÷ 25 ; 250 ;

$$25 \overline{)\, 6250} \quad \begin{array}{r} 250 \\ \hline 50 \\ 125 \\ 125 \end{array}$$

57. 32
58. 0.25
59. 4900
60. 45
61. 15.6
62. 2.5
63. 4.34 ; 12.05
64. 4.2 + (5 x 1.6) ; 4.2 + 8 ; 12.2
65. (67.5 ÷ 0.05) ÷ 0.2 ; 1350 ÷ 0.2 ; 6750
66. 53.9 – (12.8 ÷ 3.2) ; 53.9 – 4 ; 49.9
67. 6.16
68. 8.08
69. 4.39

6 Fractions

1. $1\frac{1}{9}$
2. $1\frac{4}{5}$
3. $1\frac{2}{3}$
4. $1\frac{19}{36}$
5. $1\frac{5}{12}$
6. 3
7. 34
8. $20\frac{1}{3}$
9. $\frac{5}{6} > \frac{3}{4} > \frac{2}{3} > \frac{7}{12}$
10. $3\frac{1}{4} > 2\frac{7}{8} > 2\frac{3}{4} > \frac{21}{8}$
11. $\frac{11}{5} > 1\frac{2}{3} > \frac{16}{15} > \frac{4}{5}$
12. $\frac{5}{2} > 2\frac{3}{10} > \frac{21}{10} > 1\frac{4}{5}$
13. $\frac{10}{16} + \frac{1}{16} ; \frac{11}{16}$
14. $\frac{14}{24} + \frac{7}{24} ; \frac{7}{8}$
15. $\frac{21}{35} + \frac{30}{35} ; 1\frac{16}{35}$
16. $\frac{5}{10} - \frac{1}{10} ; \frac{2}{5}$
17. $\frac{14}{15} - \frac{5}{15} ; \frac{3}{5}$
18. $1\frac{8}{18} + 8\frac{1}{18} ; 9\frac{1}{2}$
19. $4\frac{7}{6} - 3\frac{4}{6} ; 1\frac{1}{2}$
20. $9\frac{34}{24} - 2\frac{21}{24} ; 7\frac{13}{24}$
21. $6\frac{2}{4} - 5\frac{1}{4} ; 1\frac{1}{4}$
22. $10\frac{9}{12} - 2\frac{2}{12} ; 8\frac{7}{12}$
23. $8\frac{11}{11} - 1\frac{7}{11} ; 7\frac{4}{11}$
24. $5\frac{16}{14} - 4\frac{7}{14} ; 1\frac{9}{14}$
25. $5\frac{49}{56} - 2\frac{48}{56} ; 3\frac{1}{56}$
26. $2\frac{1}{15} + 2\frac{9}{15} ; 4\frac{2}{3}$
27. $7\frac{4}{4} - 7\frac{3}{4} ; \frac{1}{4}$
28. $7\frac{12}{15} + 2\frac{10}{15} ; 10\frac{7}{15}$
29. $7\frac{1}{2}$
30. $4\frac{1}{2}$
31. 12
32. 38
33. $\frac{5}{14}$
34. $\frac{4}{7}$
35. 2
36. 9
37. $6\frac{7}{8}$
38. 4
39. $6\frac{9}{10}$
40. 3
41. $3\frac{11}{25}$
42. $\frac{5}{8}$
43. $\frac{38}{47}$
44. 1
45. 9
46. $9\frac{4}{27}$
47. $12\frac{2}{5} - 2 ; 10\frac{2}{5}$
48. $7\frac{3}{8} \times 2\frac{2}{3} ; 19\frac{2}{3}$
49. $1\frac{5}{12} + 1\frac{5}{6} ; 3\frac{1}{4}$
50. 3 – 3 ; 0

ISBN: 978-1-897164-21-1

51. $3\frac{1}{2} + 6\frac{7}{8}$; $10\frac{3}{8}$

52. $5\frac{2}{5} - \frac{1}{2}$; $4\frac{9}{10}$

53. $1\frac{1}{10} + 1\frac{1}{2}$; $2\frac{3}{5}$

54. $10 - 4$; 6

55. $1\frac{4}{5} \times 1\frac{2}{5} \div 1\frac{1}{5}$; $2\frac{1}{10}$

56. $\frac{1}{2} + 3\frac{5}{6} \div 3$; $1\frac{7}{9}$

57a. $\frac{1}{3}$; 60 ; 20 ; 20

b. $\frac{1}{4}$; 60 ; 15 ; 15

c. 12 ; $\frac{2}{3}$; 8

d. 120 ; $\frac{3}{5}$; 72

e. 15 ; $1\frac{1}{2}$; $22\frac{1}{2}$

f. 20 ; $\frac{2}{5}$; 8

58. $32 \div \frac{1}{4}$; 128 ; 128

59. $6 \div \frac{3}{4}$; 8 ; 8

60. $(\frac{1}{3} + \frac{1}{4}) \times 2$; $1\frac{1}{6}$; $1\frac{1}{6}$

61.

No. of eggs : 6

Red onion : $\frac{9}{20}$ of an onion

Amount of mayonnaise: $1\frac{1}{8}$ tablespoons

Slices of bread: 6

$8 \times \frac{3}{4} = 6$

$\frac{3}{5} \times \frac{3}{4} = \frac{9}{20}$

$1\frac{1}{2} \times \frac{3}{4} = 1\frac{1}{8}$

$8 \times \frac{3}{4} = 6$

7 Ratios and Rates

1. 1.25

2. 35 words/min

3. 93 pages/day

4. $0.68/mL

5. 96 km/h

6. 0.3 m/s

7. 23 turns/s

8a. $15/h ; $25/h ; $18/h

b. Mary

9a. $20/h ; $12/h ; $18/h

b. Michael

10. ($62.19 for 3 kg) ; 20.73 ; 23.94

11. ($20.16 for 3 bags) ; 8.59 ; 6.72

12. ($2.76 for 2.4 m) ; 1.15 ; 1.54

13. ($15 for 0.5 L) ; 0.03 ; 0.05

14.-15. (Suggested answers)

14a. 1:2 ; $\frac{1}{2}$

b. 1:1 ; 1 to 1

c. 2:5 ; $\frac{2}{5}$

15a. 1:3 ; 1 to 3

b. 1:4 ; $\frac{1}{4}$

c. 3:4 ; 3 to 4

16. 17 ; 11:17 ; 4:17 ; 2:17

17. 18 ; 2:3 ; 1:6 ; 1:6

18. 15 ; 2:3 ; 2:15 ; 1:5

19. 17 ; 15:17 ; 1:17 ; 1:17

20. Bears and Raiders

21a. 5:6

b. 1:9

8 Percents

1. 65 ; 0.065 ; 6.5%

2. $\frac{88}{100} = \frac{22}{25}$; $0.88 \times 100\% = 88\%$

3. $35 \div 100 = \frac{35}{100} = \frac{7}{20}$; $35 \div 100 = 0.35$

4. $2\frac{16}{100} = 2.16$; $2\frac{4}{25} \times 100\% = 216\%$

5. 85% , 0.53 , $\frac{9}{25}$

6. 0.73 , $\frac{29}{40}$, 70.5%

7. $1\frac{1}{2}$, 123% , 1.15

8. $15 \times 30\% = $4.5

9. 400 cm $\times$ 25% = 100 cm

10. 20 kg $\times$ 145% = 29 kg

11. 0.075 x 80 L = 6 L

12. $\frac{15}{24} \times 100\% = 62.5\%$

13. $\frac{8}{2.5} \times 100\% = 320\%$

14. 120 = a number x 80% ; a number = $120 \div 0.8 = 150$

15. 48 = a number x 60% ; a number = $48 \div 0.6 = 80$

16. 4

17. 20.6

18. 150.4

19. 16

20. 100

21. 72

22. 54

23. 30

24a. 5.65

b. 50.85

25a. 8.48

b. 64.98

9 Equations

1. $4n$

2. $\frac{n}{8}$

3. $\frac{n}{2}$

4. $\frac{n-7}{2}$

5. $2n + 3$

6. $\frac{4}{n-5}$

7. $0.35(n-2)$

8. $n^2 + 7$

9.-12. (Suggested answers)

9. 9 divided by a number

10. The product of a number and 5

11. The product of a number and 4; then increased by 1

12. The product of 6 and a number decreased by 2

13. 3 ; -2 ; 18 + 6 ; 24

14. 3 ; -2 ; 20 + 42 ; 62

15. -2 ; 1 ; 6 + 2 ; 8

16. 1 ; 3 ; 8 - 2 (9) ; -10

17. x^2: $2x^2, -2x^2, 9x^2, x^2$

xy: $5xy, xy, -2xy, 3xy$

constant: 3 , -4 , 8 , 0

18. $+5, +2x^2, -6x, +3xy$; 4

19. $+3x^2, +6xy, -3x, +y^2, +9$; 5

20. $+h^2, -h$; 2

21. $-4e, -9e^2, +7e^3, +2$; 4

22. $+123abc$; 1

23. $3x + 8y$

24. ($3x^2$) $- 4$ ($+ 5x^2$) $+ x$; $8x^2 + x - 4$

25. $\boxed{8}$ ($- 2u^2$) $+ 3u$ ($+ 5u^2$) $\boxed{+ 1}$; $3u^2 + 3u + 9$

26. (k^2) $\boxed{+ 4k}$ ($- 6k^2$) $+ 3$ $\boxed{- 2k}$; $-5k^2 + 2k + 3$

27. (t^2) $- 4i$ ($+ 4t^2$) ; $5t^2 - 4i$

28. ($7x^2y$) $\boxed{- 6xy^2}$ ($- 8x^2y$) $\boxed{+ 14xy^2}$; $-x^2y + 8xy^2$

29. $4a + 4b$

30. $2(c + d)$

$2c + 2d$

31. $m(m + n)$

$m^2 + mn$

32. $3x(x + 8)$

$3x^2 + 24x$

33. $5x(y + 2)$

$5xy + 10x$

34. $3(2x^2 - 6)$

$6x^2 - 18$

35. $9b(b - a + 4)$

$9b^2 - 9ab + 36b$

36. $5(6x - 7 + x^2)$

$30x - 35 + 5x^2$

37. $4(-9 + 6m + n)$

$-36 + 24m + 4n$

38. $5(7x^2 - 6)$; $35x^2 - 30$

ISBN: 978-1-897164-21-1

39. $e(-e+1)$; $-e^2 + e$

40. $3(-2y + 9y^2)$; $-6y + 27y^2$

41. $4h(-h^2 - 2h)$; $-4h^3 - 8h^2$

42. 15

43. $n + 4 - 4 = 24 - 4$; $n = 20$

44. $s - 16 + 16 = 2 + 16$; $s = 18$

45. $4i \div 4 = 12 \div 4$; $i = 3$

46. $\frac{k}{9} \times 9 = 10 \times 9$; $k = 90$

47. $14n \div 14 = 28 \div 14$; $n = 2$

48. $\frac{3}{4}y \div \frac{3}{4} = 24 \div \frac{3}{4}$; $y = 24 \times \frac{4}{3}$; $y = 32$

49. $\frac{4}{3}e \div \frac{4}{3} = 32 \div \frac{4}{3}$; $e = 32 \times \frac{3}{4}$; $e = 24$

50. $8x = 4.8$; $8x \div 8 = 4.8 \div 8$; $x = 0.6$

51. $2n = 24$; $2n \div 2 = 24 \div 2$; $n = 12$

52. $11a = 99$; $11a \div 11 = 99 \div 11$; $a = 9$

53. $\frac{i + 2i}{3} = 4$; $i = 4$

54. $(w)(11) = 165$; $11w \div 11 = 165 \div 11$; $w = 15$; 15

55. $2y + 5 = 39$; $y = 17$; There are 17 marbles in each box.

56. $p \times 15\% = 45$; $p = 300$; The cost of the ring is \$300.

Review

1. C

2. B

3. C

4. B

5. A

6. D

7. D

8. B

9. C

10. +2

11. +14

12. +18

13. -3

14. -29

15. +4

16. 3.55

17. 2.75

18. 10.84

19. 2.88

20. $= 1\frac{1}{3} \times \frac{9}{8} = 1\frac{1}{2}$

21. $= 7\frac{2}{7} - 6\frac{3}{7} = \frac{6}{7}$

22. $= 1\frac{7}{18} - \frac{1}{18} = 1\frac{1}{3}$

23a. 5 ; 25

b. 4 ; 20

c. 1:4

24. 67.2

25. 60

26. 15%

27. 80

28. 125

29. $4x^2$; 1

30. $5a + 3b$; 2

31. $7k^2 + 2k$; 2

32. $-7m^3 + 7m^2 - 7m$; 3

33. $5 + n - 5 = 17 - 5$; $n = 12$

34. $\frac{16x}{16} = \frac{80}{16}$; $x = 5$

35. $\frac{k}{4} \times 4 = 12 \times 4$; $k = 48$

36. $\frac{7m}{7} = \frac{56}{7}$; $m = 8$

37. $\frac{21y}{21} = \frac{63}{21}$; $y = 3$

38. $\frac{12e}{12} = \frac{156}{12}$; $e = 13$

39. $584 \times 75\% = 438$; There were 438 female customers.

40. $\frac{c}{2} - 3 = 6$; $\frac{c}{2} - 3 + 3 = 6 + 3$; $\frac{c}{2} = 9$; $c = 18$;

There are 18 candies in the bag.

41. $5l = 225$; $\frac{5l}{5} = \frac{225}{5}$; $l = 45$;

The height of the parallelogram is 45 cm.

ISBN: 978-1-897164-21-1

1 Number Theory

1. 2^3 — Exponent $= 3$ — Base $= 2$
2. 7^5 — Exponent $= 5$ — Base $= 7$
3. 9^6 — Exponent $= 6$ — Base $= 9$
4. 5^8 — Exponent $= 8$ — Base $= 5$
5. 625
6. 1
7. 9
8. 144
9. 64
10. 1
11. 13
12. 1
13. 216
14. 10^2
15. 10^5
16. 10^3
17. 10^4
18. 10^6
19. 10^5
20. 10^7
21. 1; 4; 6; 3
22. $3 \times 10^3 + 7 \times 10^1 + 5 \times 10^0$
23. $1 \times 10^2 + 5 \times 10^1 + 9 \times 10^0$
24. $4 \times 10^4 + 2 \times 10^3 + 6 \times 10^1 + 2 \times 10^0$
25. 21 432
26. 30 509
27. 600 332
28. 40 305

29.
$2 \times 2 \times 3$
3

30.
$2 \times 2 \times 3 \times 3$
$2^2 \times 3^2$

31.
$2 \times 2 \times 7 \times 7$
$2^2 \times 7^2$

32.
$5 \times 5 \times 3 \times 3$
$5^2 \times 3^2$

33. 4
34. 2
35. 2^6
36. 3×5^2
37. $2^4 \times 5^2$
38. $2^3 \times 3 \times 5$
39. $2^2 \times 11^2$
40. $2^3 \times 3^2$
41. 3×5^3
42. $2^3 \times 5^2$
43. $2 \times 2 \times 2 \times 2$
   4
44. $3 \times 3 \times 7 \times 7$
   21
45. $2 \times 2 \times 5 \times 5$
   10
46. $2 \times 2 \times 11 \times 11$
   22
47. 16
48. 30
49. 12
50. 18

Activity

1. 25
2. 49

2 Algebraic Expressions

1. a. $8k + 4$ — b. $8 \times 7 + 4 = 60$
2. a. $12m + 7$ — b. $12 \times 5 + 7 = 67$
3. a. $5p + 3$ — b. $5 \times 8 + 3 = 43$
4. 27
5. 90
6. 16
7. 75
8. 22
9. 36
10. 10
11. 7
12. 16
13. 9
14. 14
15. 8.7
16. 3.6
17. 0.8
18. 0.4
19. 6.3
20. 0
21. 1.6
22. 36
23. $3\frac{1}{2}$
24. $\frac{3}{4}$
25. $\frac{2}{5}$
26. 10
27. 15
28. 22
29. 5
30. 1
31. 3.5
32. 0
33. 0.25
34. 25
35. 1.2

36. a.

y	(x, y)
2	(0, 2)
4	(2, 4)
5	(3, 5)

b.
$y = x + 2$

c. 3

37. a.

y	(x, y)
0	(1, 0)
2	(3, 2)
4	(5, 4)

b.
$y = x - 1$

c. 3

38. a.

y	(x, y)
6	(0, 6)
4	(2, 4)
2	(4, 2)

b.
$y = 6 - x$

c. 3

39. a.

y	(x, y)
1	(2, 1)
2	(4, 2)
3	(6, 3)

b.
$y = x \div 2$

c. 4

Activity

1. 50; (1, 50)
   100; (2, 100)
   200; (4, 200)
   300; (6, 300)
2. $y = 50x$

ISBN: 978-1-897164-21-1

3 Fractions

1. $3\frac{1}{2}$
2. 2
3. $21\frac{1}{4}$
4. $14\frac{1}{2}$
5. 48
6. 33
7. 30
8. $7\frac{7}{8}$
9. $1\frac{3}{4}$
10. 10
11. 4
12. $\frac{1}{6}$
13. $2\frac{2}{11}$
14. $4\frac{1}{6}$
15. 22
16. $\frac{7}{18}$
17. $2\frac{1}{3}$
18. $\frac{5}{8}$
19. $3\frac{3}{4}$
20. 4
21. $\frac{3}{7}$
22. $1\frac{2}{3}$
23. $1\frac{1}{2}$
24. 7
25. $\frac{1}{8}$
26. $\frac{5}{7}$
27. $\frac{1}{3}$
28. $\frac{4}{5}$
29. $\frac{6}{11}$
30. 6

31. climbing

32. $7\frac{1}{2} \div 3\frac{1}{3} = 2\frac{1}{4}$

   $2\frac{1}{4}$

33. $1\frac{1}{5} \times 3\frac{1}{3} = 4$

   4

34. $2\frac{1}{2} \times \frac{2}{3} = 1\frac{2}{3}$

   $1\frac{2}{3}$

35. $\frac{3}{4} \div 4 = \frac{3}{16}$; $\frac{3}{16} \times 1000 = 187\frac{1}{2}$

   $187\frac{1}{2}$

36. $\frac{7}{8} - \frac{7}{8} \times \frac{2}{5} = \frac{21}{40}$

   $\frac{21}{40}$

Activity

1. $12\frac{3}{8}$
2. $18\frac{9}{16}$
3. $1\frac{1}{2}$

4 Percents

1. 36%
2. 62.5%
3. 42.86%
4. 68.75%
5. 180%
6. 285.71%
7. 126.67%

8. $\frac{1}{2}$
9. $\frac{23}{50}$
10. $\frac{6}{25}$
11. $\frac{2}{25}$
12. $\frac{1}{50}$
13. $\frac{39}{50}$
14. 45
15. 38.89
16. 37.5
17. 41.67
18. 37.5
19. 33.6
20. 162.5
21. 74.8
22. 24
23. 16.2
24. a. 18 b. 24 c. 78
25. 10
26. 30
27. 3.75
28. 5

29. $\frac{65}{250} \times 100\% = 26\%$

   26

30. 35.28
31. 62.06
32. 14.34
33. 62.99
34. 21.39
35. 24.15
36. 64.40
37. 26.44
38. 48.98
39. 44.03
40. 35%; $\$31.50$
41. $\$75$; 15%
42. 20%; $\$28.68$
43. 7%; $\$42.80$
44. $\$29.50$; 12%
45. 6%; $\$4.95$
46. 22.95
47. 35.55
48. 36.11

Activity

37.8

5 Measurement

1. 28; 40
2. 18; 20.25
3. 9.7; 2.75
4. 13.6; 3.96
5. 29; 30
6. 26.2; 15
7. 10.5
8. 8.85
9. 19
10. 35.85
11. 2.34
12. 393
13. 72
14. $140\ 000$
15. $165\ 000$
16. $84\ 500$
17. $9\ 400\ 000$
18. 126
19. 4.732
20. $4 \times 3.8 \times 49 = 744.80$

   744.80
21. $(4 \times 2.5 \times 2 + 3.8 \times 2.5 \times 2) \div 24 = 1.625$

   2
22.
23. ✓
24. ✓
25.
26. 24
27. 360
28. 97.5
29. 324
30. 695.4
31. 12; 43.2
32. 17.5; 143.5
33. 10.5; 63
34. 6; 57
35. 6.3; 37.8
36. 18.36; 91.8
37. 27; 248.4
38. 2.356
39. 1.73
40. 0.36
41. $1\ 500\ 000$
42. $7\ 000$
43. $600\ 000$
44. $1\ 040\ 000$
45. $120\ 000$; 0.12
46. 9.6

ISBN: 978-1-897164-21-1

Activity

1. 6; 3; 0.5 x 6 - 1 + 3 = 5
2. 6; 4; 0.5 x 6 - 1 + 4 = 6

6 Approximation

1. A
2. E
3. A
4. E
5. A
6. 14.5
7. 9.5
8. 13.5
9. 13.5
10. 2 cm; 2.5 cm
11. 2 cm; 2 cm
12. 3 cm; 3.5 cm
13. 10 cm; 10 cm

Activity

Estimate: Individual answer
Exact: 30

7 Integers

1. a. +$1 200   b. -$450
2. a. +1 kg   b. -3 kg
3. <
4. <
5. <
6. >
7. >
8. <
9. -6, -3, 0, +2, +4
10. -4, -3, +1, +5, +7
11. -5, -1, 0, +3, +6
12. (+5) + (-8) = -3
13. (-2) + (+6) = +4
14. (-3) + (-4) = -7
15. (+6) + (+2) = +8
16. (+5) + (-5) = 0
17. -12
18. -5
19. -3
20. +12
21. 0
22. -4
23. +11
24. -5
25. -7
26. +4
27. +8
28. -6
29. -6
30. +7
31. -26
32. +20
33. +3
34. -4
35. +9
36. -10
37. (-5) + (+4)   38. (+4) + (-7)
    -1   -3
39. (-13) + (-2)   40. (+6) + (+10)
    -15   +16
41. +2
42. +11
43. +8
44. +22
45. -5
46. -4
47. +7
48. -8
49. -3
50. +1
51. +22
52. -54
53. -30
54. -5
55. +20
56. +5
57. -7
58. +3
59. -35
60. -20
61. Practice makes perfect.

Activity

1. Sunday
2. Tuesday
3. 9

8 Fractions, Decimals, and Percents

1. 20%; $\frac{1}{5}$
2. 0.28; $\frac{7}{25}$
3. 45%; 0.45
4. 65%; $\frac{13}{20}$
5. 66.67%; 0.67
6. 140%; $1\frac{2}{5}$
7. 270%; 2.7
8. Dave; 1; Tony
9. Helen; 0.15; Elaine
10. 3; math; history
11. 11
12. 15
13. 4
14. 20
15. 3
16. 13.2
17. 18.9
18. 0.175
19. 0.74
20.

Dave White		MATH TEST	
1. 2^2 x 0.3 - 1	= 0.2 ✓	2. $(14 - \frac{1}{2} \times 10)^2$	= 81 ✓
3. 7 - 0.65 x 10	= 63.5 ☐	4. $18 - \frac{1}{4} \times 12 \div 2$	= 7.5 ☐
5. 30% x $(15 - 9)^2$	= 10.8 ✓	6. $(12 - 2) \times 6 - 4^2$	= 44 ✓
7. $\frac{7}{12}$ x 144 - 6 ÷ 3	= 82 ✓	GRADE:	$\frac{6}{8}$ = 75 %
8. 10 - (35% x 9 - 2)	= 8.85 ✓		

21.

Steve Lindsay		MATH TEST	
1. 3^2 x (15 - 12)	= 27 ✓	2. 20 + 6 x 3 ÷ 2	= 29 ✓
3. $(4 + 5)^2 - 10 \times \frac{1}{2}$	= 76 ✓	4. (40 - 26) ÷ 2 - 7	= 0 ✓
5. (0.2 + 50%) x 6^2	= 25.2 ✓	6. $(0.5 + 5) \times 2^2$	= 121 ☐
7. $\frac{2}{5}$ x 45 - 16 x 80%	= 5.2 ✓	GRADE:	$\frac{6}{8}$ = 75 %
8. (25 - 9) ÷ 4 + 50% x 6	= 27 ☐		

22. a. 547.25   b. 584
23. a. 9.03   b. 10.19
24. a. 325   b. 220
25. a. 24   b. 34.60

Activity

1. 5 x 2 + 4 - 3 - 1 = 10
2. 3 x (2 + 1) + 5 - 4 = 10
3. 5 x 2 x 1 x (4 - 3) = 10
4. ((4 x 2) - 5) x 3 + 1 = 10

Midway Test

1. 8^{12} Exponent = 12   Base = 8
2. 4^5 Exponent = 5   Base = 4
3. 3^4; 9
4. 2^4; 4
5. 2^6; 8
6. 2^4 x 5^2; 20
7. 2^2 x 7^2; 14
8. 2^2 x 11^2; 22

9. 4
10. 21
11. 8.8
12. 1.3
13. 25
14. 5.4
15. 1.2
16. 2
17. $1\frac{1}{3}$
18. $15\frac{1}{2}$
19. 3
20. $\frac{3}{10}$
21. $\frac{4}{5}$
22. 2
23. $\frac{1}{2}$
24. $\frac{1}{3}$
25. 8
26. $\frac{7}{12}$
27. 104
28. 41.67
29. 25
30. $10; $30
31. $22.50; $52.50
32. $126; 20%
33. 15%; $75.82
34. $4.32; $40.32
35. 8%; $16.74
36. $75.60; 15%
37. $31\frac{1}{2}$; $40\frac{5}{8}$
38. 20.8; 20
39. 33.8; 43.7
40. 15.8; 16.12
41. 28.8
42. 2
43. 29.5
44. 72
45. 0.52
46. 18 500
47. 4.56
48. 850 000
49. 5 cm; 5 cm
50. 8 cm; 8.5 cm
51. 13 cm; 12.5 cm
52. -11
53. $-21 \div -3 = 7$
54. $4 - 12 = -8$
55. $-12 \times 6 = -72$
56. $-6 \times 3 = -18$
57. $-12 \div -6 = 2$
58. $-21 + 3 - (-12) = -6$
59. $5 - (-6 + 14) = -3$
60. $y + 12 = +5$
    $y = -7$
    -7
61. $x = 7 - 11$
    $x = -4$
    -4
62. $1\frac{3}{5}$; 160%
63. 0.45; $\frac{9}{20}$
64. Dave's savings = $70 x 38.5% = $26.95
    Steve's savings = $50 x 46.3% = $23.15
    Dave; 3.80; Steve
65. a. 80 x 85% = 68
       68
    b. $\frac{72}{80}$ x 100% = 90%
       90
66. a. 33
    b. 0.50

9   Coordinates

1. (0, 10)
2. (6, 10)
3. (8, 10)
4. (14, 10)
5. (1, 8)
6. (3, 8)
7. (16, 8)
8. (2, 7)
9. (4, 7)
10. (5, 5)
11. (11, 5)
12. (18, 0)
13. I
14. C
15. G
16. K
17. (-6, -3)
18. (0, -3)

19. (4, 3)
20. (5, 5)
21. (-4, -2)
22. (3, -5)
23. (-3, 5)
24. (-4, 2)
25. B, C, F, H
26. A, D, G
27. L, N, R
28. M, Q, S
29. I, J, K
30. E, J, P
31. B(-6, 3), C(-6, -5), D(-8, -5)
32. E(2, 3), F(5, 3), G(7, -1), H(2, -1)
33. L(-3, -1), M(-3, -5), N(1, -5)
34. a. 2
    b. 8
    c. 16
35. a. 3
    b. 5
    c. 4
    d. 16
36. a. 4
    b. 4
    c. 8
37. a.

y	-1	2	4
(x, y)	(0, -1)	(3, 2)	(5, 4)

b. 3
c. 2

38. a.

y	-3	1	5
(x, y)	(-1, -3)	(1, 1)	(3, 5)

b. 3
c. 0

Activity

1. $y = x + 2$
2. $x = 1$
3. (1, 3)

10   More about Algebraic Expressions

1. $y^2 - 25$
2. $(25 + y) \div 5$
3. $25 + y^2$
4. $(y - 25)^2$ or $(25 - y)^2$
5. $(25 + y) \times 2$
6. $25y - 25$
7. 24
8. 42
9. 36
10. 0.9
11. 12
12. -11
13. 0
14. 8

ISBN: 978-1-897164-21-1

15. $5x^2$, $3x$, $2y$, -8; 4

16. $3a$, $2b$, -6; 3

17. $0.5m$, $2n$, $-2p$, 9; 4

18. $8u^3$, $-4u^2$, 3; 3

19. like

20. unlike

21. like

22. unlike

23. $10p + 4$

24. $6d + 13$

25. $-9k - 9$

26. $1\frac{1}{4}m - 7$

27. $15n^2 - 10n + 5$

28. $-6a + 3b + 4$

29. $9m - 4n + 5$

30. $-4pq$

31. $5x^2y + 2y$

32. $3x + 9$

33. $3y - 2$

34. $8h - 12$

35. $-2m - 4$

36. $4n + 12$

37. $5p + 8$

38. 15

39. 7

40. 36

41. 0

42. -8

43. 11

44. 8

45. -1

46. 15.6

47. -2.2

48. 6.6

49. 0.2

50. a. 18   b. 60   c. 16

    d. 384

51. a. 45   b. 36   c. 360

    d. 10

52. a. 60   b. 100   c. 14

    d. 0.3

53. a. 87.5%   b. 32.5%   c. 93.75%

    d. 80%

54. $8y = 16$

$8y \div 8 = 16 \div 8$

       $y = 2$

Check: $3(2) + 5(2) = 16$

55. $x - 3 = 6$

$x - 3 + 3 = 6 + 3$

       $x = 9$

Check: $(9) - 7 + 4 = 6$

56. $2m = -12$

$2m \div 2 = -12 \div 2$

       $m = -6$

Check: $8(-6) - 6(-6) = -12$

57. $2a = 14$

$2a \div 2 = 14 \div 2$

       $a = 7$

Check: $8(7) - 12(7) + 6(7) = 14$

58. $3y + 14 = 38$

$3y + 14 - 14 = 38 - 14$

       $3y = 24$

    $3y \div 3 = 24 \div 3$

         $y = 8$

8

59. $4 \times m + 20 = 100$

$4m + 20 - 20 = 100 - 20$

      $4m = 80$

   $4m \div 4 = 80 \div 4$

       $m = 20$

   20

60. $8 \times 5 \times p = 500$

     $40p = 500$

$40p \div 40 = 500 \div 40$

       $p = 12.50$

   12.50

Activity

10	⑦	4
16	5	⑭
⑥	12	8

11   Angles and Lines

1. 30°; 90°; adjacent

2. 30°; 30°; opposite

3. 30°; 60°; complementary

4. 30°; 150°; supplementary

5. 37°

6. 34°

7. 95°

8. 53°

9. 45°; 80°

10. 30°; 90°

11. 45°; 55°

12. 30°

13. 110°

14. 67°

15. 156°; 24°

16. 100°; 35°

17. 78°; 102°

18. corresponding

19. alternate

20. alternate

21. corresponding

22. 120°; 60°

23. 100°; 80°

24. 125°; 125°

25. 54°; 126°

26. 135°; 135°

27. 145°; 35°

28. 61°; 119°

29. 25°; 45°

30. 40°; 45°

31. 122°; 58°

32. 80°; 42°

33. 60°; 65°

34. 60°; 96°

35. 124°; 56°

36. 58°; 50°

Activity

1. AB

2. same

ISBN: 978-1-897164-21-1

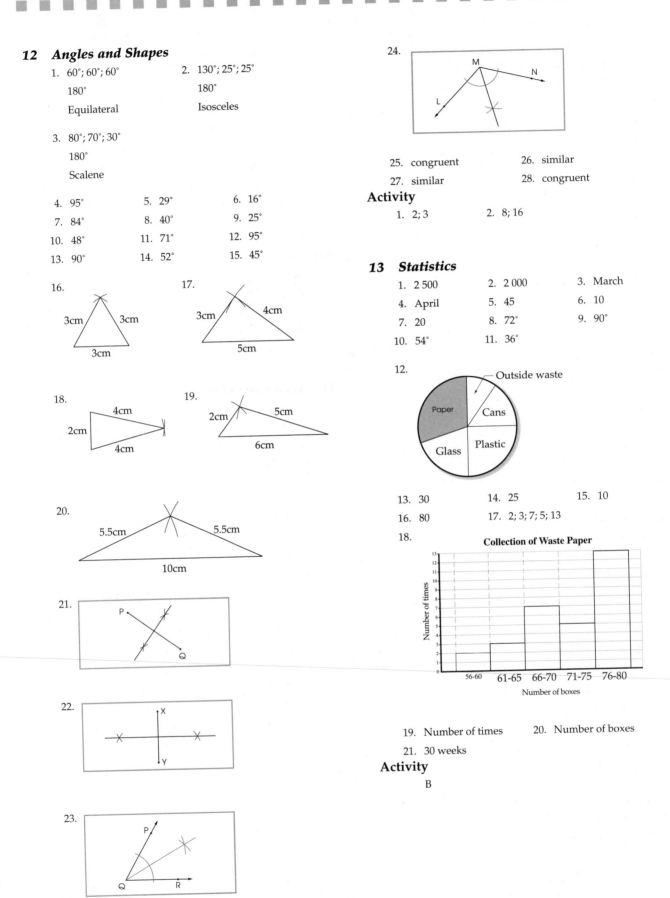

12 Angles and Shapes

1. 60°; 60°; 60°
    180°
    Equilateral

2. 130°; 25°; 25°
    180°
    Isosceles

3. 80°; 70°; 30°
    180°
    Scalene

4. 95°
5. 29°
6. 16°
7. 84°
8. 40°
9. 25°
10. 48°
11. 71°
12. 95°
13. 90°
14. 52°
15. 45°

16.
17.
18.
19.
20.
21.
22.
23.
24.

25. congruent
26. similar
27. similar
28. congruent

Activity
1. 2; 3
2. 8; 16

13 Statistics

1. 2 500
2. 2 000
3. March
4. April
5. 45
6. 10
7. 20
8. 72°
9. 90°
10. 54°
11. 36°

12.

13. 30
14. 25
15. 10
16. 80
17. 2; 3; 7; 5; 13

18.

19. Number of times
20. Number of boxes
21. 30 weeks

Activity
    B

ISBN: 978-1-897164-21-1

14 Transformations

1.

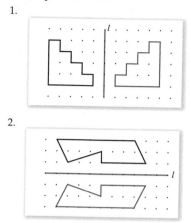

2.

3. 5 units right and 2 units up
4. 5 units left and 3 units down
5. 3 units down
6. 7 units right and 3 units down
7. (5,4); 90°, counter clockwise
8. (6, 1); 90°, counter clockwise
9. (11, 3); 180°, counter clockwise or clockwise

Activity
4; 8; 5; 6

15 Probability

1. $\frac{2}{8}$ ($\frac{1}{4}$) 2. $\frac{2}{8}$ ($\frac{1}{4}$)
3. $\frac{3}{8}$ 4. $\frac{2}{8}$ ($\frac{1}{4}$)
5. $\frac{3}{8}$ 6. $\frac{1}{8}$
7. $\frac{3}{8}$ 8. $\frac{2}{8}$ ($\frac{1}{4}$)
9. $\frac{1}{8}$ 10. $\frac{2}{8}$ ($\frac{1}{4}$)
11. $\frac{5}{8}$ 12. $\frac{4}{8}$ ($\frac{1}{2}$)
13. $\frac{7}{8}$
14. a. 9 b. $\frac{3}{9}$ ($\frac{1}{3}$)
    c. $\frac{4}{9}$
15. a. 12 b. $\frac{1}{12}$
    c. $\frac{1}{12}$ d. $\frac{2}{12}$ ($\frac{1}{6}$)
    e. $\frac{2}{12}$ ($\frac{1}{6}$)

Activity
1. A $\frac{3}{8}$ B $\frac{1}{4}$
   C $\frac{1}{5}$
2. A

Final Test
1. A(-8, 3), B(-8, -4), C(-3, -4)
2. R(4, 3), S(4, -1), T(9, -1), U(9, 3)
3. 17.5 4. 20
5. B, C 6. S, T
7. P 8. L
9. $4 + x^2$ 10. $4y - 16$
11. $9m + 3$ 12. $(q - 5) \div 4$ or $(5 - q) \div 4$
13. 18 14. 36
15. -42 16. 45
17. $3a + b$ 18. $16 + 3m$
19. $3n + 33$ 20. $-4t + 15$
21. $2.5a = 9$ 22. $3.5b = 31.5$
    $a = 3.6$ $b = 9$
23. 26°; 26° 24. 45°
25. 60°; 60° 26. 50°
27. 72° 28. 54°
29. 94°; 86° 30. 65°; 115°
31. 72°; 108° 32. 30°; 150°
33. similar 34. congruent
35. congruent 36. 101°
37. 24° 38. 45°
39. 79°
40.

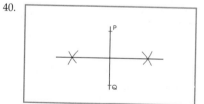

41.

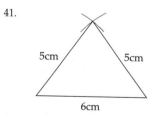

5cm 5cm
6cm

42.

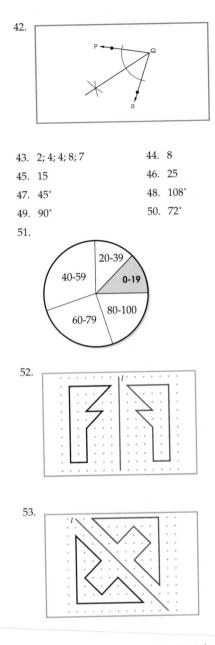

65. a. HHH, HHT, HTH, HTT, THH, TTH, THT, TTT

    b. $\frac{3}{8}$ c. $\frac{4}{8}$ ($\frac{1}{2}$)

43. 2; 4; 4; 8; 7   44. 8

45. 15   46. 25

47. 45°   48. 108°

49. 90°   50. 72°

51.

54. 7; left; 2; up   55. 10; right; 1; down

56. (5, 2); 90° counter clockwise

57. (9, 3); 90° clockwise

58. $\frac{3}{10}$ 59. $\frac{2}{10}$ ($\frac{1}{5}$)

60. $\frac{2}{10}$ ($\frac{1}{5}$)   61. 0

62. $\frac{4}{10}$ ($\frac{2}{5}$)   63. $\frac{4}{10}$ ($\frac{2}{5}$)

64. a. AA, AB, AC, AD, BB, BA, BC, BD, CA, CB, CC, CD, DA,

       DB, DC, DD

    b. $\frac{4}{16}$ ($\frac{1}{4}$)   c. $\frac{7}{16}$

ISBN: 978-1-897164-21-1

Unit 1

1. $3 \times (6 - 1) = 15$ The final cost of her purchase is 15 dollars.
2. $(3 \times 19 - 5) \div 2 = 26$ Joy pays $26.00.
3. $10 + 12 \div 2 = 22 \div 2 = 11$ It gives an answer 11. It is wrong.
4. It doesn't follow the order of operations.
5. $10 + 12 \div 2 = 10 + 6 = 16$ The correct answer is 16.
6. No. $(2^2 + 3^2) = (4 + 9) = 13$; $(2+3)^2 = 5^2 = 25$
7. $(6 + 5) \times 7 + 4 \times 10 + 20 = 137$ She earned $137.00.
8. $(10 + 3) \times 2 = 26$; $10 \times 2 + 3 \times 2 = 26$
9. $(35 + 3) \div 19 + 7 \times (22 - 13) = 65$

Challenge

Suggested answer: $200 = 2 \times (1 + 9) \times (4 + 6) \times (5 - 3 + 7 - 8)$

Unit 2

1. Suggested answer:
   $(25, 15)$; $25 + 15 \neq 46$ ✗ ; $(26, 16)$: $26 + 16 \neq 46$ ✗
   $(27, 17)$; $27 + 17 \neq 46$ ✗ ; $(28, 18)$: $28 + 18 = 46$ ✔
   The long call lasted 28 minutes; the short call lasted 18 minutes.
2. Suggested answer:
   The total amount: $100 - 25 = 75$
   $20 + 55 = 75$, $55 - 20 = 35$ ✗
   $30 + 45 = 75$, $45 - 30 = 15$ ✔
   The price of the more expensive one is $ 45.00
3. $933 \div 2 = 466.5$, $466 < 466.5 < 467$
   The two numbers are 466 and 467.
4. Suggested answer:
   $11^2 - 10^2 = 121 - 100 = 21$ ✗
   $12^2 - 11^2 = 144 - 121 = 23$ ✗
   $13^2 - 12^2 = 169 - 144 = 25$ ✔
   The numbers are 12 and 13.
5. If a 2-digit number has a remainder of 2 when divided by 3, this number cannot be a multiple of 3. Also the ones digit of this number must be 3 or 8 if it has a remainder of 3 when divided by 5. The numbers are 23, 38, 53, 68, 83 and 98.
6. Suggested answer:

width(m)	10	15	20	25	30
length(m)	40	35	30	25	20
area(m²)	400	525	600	625	600

The largest possible straight-sided area is 625 m².
7. Suggested answer:

Total cost	each of 20 paid	each of 25 paid	difference
400	20	16	4 ✗
500	25	20	5 ✔

The total cost is $500.00
8. $(106 - 55) \div 3 + 5 = 22$
   The original price of each T-shirt is $22.00.
9. Difference: $4000 - 1200 - 1800 = 1000$
   The difference in altitude between the two planes is 1000 m.
10. a. $4 + 29 \times 3 = 91$ The total length of the fence is 91 m.
    b. $4 + 39 \times 3 = 121$ The total length of the fence is 121 m.
11. a. Distance : $(42 - 2) \div 2 = 20$ He has run 20 km already.
    b. Speed : $20 \div 2 = 10$ His average speed is 10 km/h.
    c. Time : $42 \div 7 = 6$ He will take 6 hours to complete the race.
    d. Difference : $(10 \times 3) - (7 \times 3) = 30 - 21 = 9$
       Ron is 9 km ahead of Fred.
    e. Speed : $(42 - 10 \times 3) \div (5 - 3) = 12 \div 2 = 6$
       The average speed is 6 km/h.
    f. Distance : $42 - 7 \times 5 = 42 - 35 = 7$ Fred is 7 km from the end.
12. a. No. of loonies : $1000000 \div 27 = 37037.037$
       It would take 37 038 loonies.
    b. No. of loonies : $1000 \div 2 = 500$
       There would be 500 loonies.
13. Time: 12 h 35 min $- 1$ h 40 min $= 10$ h 55 min   It was 10:55 am.
14. Minutes : $365 \times 24 \times 60 = 525600$
    There will be 525 600 minutes in the year 2002.
15. No. 2002 has 365 days but 2004 is a leap year. It has 366 days.
16. a. No. of students : $200 - 160 = 40$
       40 Grade 7 students do not have brown eyes.
    b. No. of students : $160 - 90 = 70$
       70 brown-eyed Grade 7 students do not have brown hair.
17. No. of times : $90 \times 365 \times 24 \times 60 \times 70 = 3311280000$
    Her heart will beat 3 311 280 000 times throughout her life.
18. $3\ 311\ 280\ 000 = 3.3 \times 10^9$
19. 190 billion $= 190\ 000\ 000\ 000 = 1.9 \times 10^{11}$
    b is correct. For a and c, the numbers 19 and 0.19 are not within 1 to 10.
20. Last number : $5 \times 10^6 \times 6 - 6 \times 10^6 \times 5 = 0$
    The last number is 0.
21. No. of kilometres : $4.6 \times 10^{10} \div 1000 \div 100 = 4.6 \times 10^5$
    There are 4.6×10^5 kilometres.
22. Capacity : $100 \times 2 \times 60 \times 2 = 2.4 \times 10^4$
    The capacity of the aquarium is 2.4×10^4 mL (24 L).
23. a. No. of games : $16 + 8 + 4 + 2 + 1 = 31$
       31 games must be played.
    b. No. of games : $32 + 16 + 8 + 4 + 2 + 1 = 63$
       63 games must be played.
    c. $64 + 63 = 127$. 127 games must be played. The no. of games is 1 less than the total number of teams.

Challenge

1. $(2 - 1) + (4 - 3) + ... + (4800 - 4799)$
   $\underbrace{\qquad\qquad\qquad}_{2400 \text{ terms}}$
   $= 1 + 1 + ... + 1$
   $= 2400$ The value is 2400.
2. Suggested answer:
   Speed $(40, 55)$, time : $300 \div 40 + 300 \div 55 = 12.95$ ✗
   Speed $(50, 65)$, time : $300 \div 50 + 300 \div 65 = 10.62$ ✗
   Speed $(60, 75)$, time : $300 \div 60 + 300 \div 75 = 9$ ✔
   The average speed is 60 km/h from Antville to Beechwood, and 75 km/h from Beechwood to Antville.

Unit 3

1.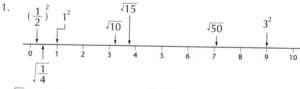

2. $\sqrt{9}$ is the square root of 9, i.e. $\sqrt{3 \times 3} = 3$
   9^2 is the square of 9, i.e. $9 \times 9 = 81$
3. a. True   b. False   c. True   d. False
4. $1, 4, 9, 16, 25, 36, 49, 64, 81, 100$
   $+3\ +5\ +7\ +9\ +11\ +13\ +15\ +17\ +19$
   Differences increase by 2.
5. Yes. $4 \times 9 = 36$ (a perfect square)
6. One side : $\sqrt{9} = 3$ Perimeter : $3 \times 4 = 12$
   Brad needs 12 m of lumber.
7. One side : $\sqrt{53000} \approx 230$ Each side is about 230 m.
8. Difference : $\sqrt{289} - \sqrt{121} = 6$ The difference is 6 m.

Challenge

Shaded area : $5 \times 5 - 3 \times 3$
$= 25 - 9$
$= 16$ Each side of the square is 4 cm.

4 cm
4 cm

Unit 4

1. 5^4 means 4 fives are multiplied, i.e. $5 \times 5 \times 5 \times 5 = 625$
   5×4 means 5 times 4, i.e. $5 \times 4 = 20$

ISBN: 978-1-897164-21-1

2. $3^4 = 3 \times 3 \times 3 \times 3 = 81$, $4^3 = 4 \times 4 \times 4 = 64$
   3^4 is bigger than 4^3.

3. $243 = 3 \times 3 \times 3 \times 3 \times 3 = 3^5$

4. There are 4 six months in 2 years.
   $100 \times 2 \times 2 \times 2 \times 2 = 100 \times 2^4 = 1600$
   There will be 1600 ducks in 2 years.

5. The amount of money that Pat puts in her piggy bank doubles every day. She will put 2^9 on Sept 10, i.e. 512¢ or $5.12.

6. One billion = 1 000 000 000 = 1×10^9 There are 9 zeros.

7. Diameter : $12800 \times 11 = 140800 \approx 1.4 \times 10^5$
   The diameter of Jupiter is about 1.4×10^5 km.

8. Distance : 93×1.6 million $= 148.8 \times 10^6 = 1.488 \times 10^8$
   The distance is about 1.488×10^8 km.

9. Population: $100000 \times 2^2 = 400000 = 4 \times 10^5$
   The population will be 4×10^5.

10. Population: $100000 \div 2 = 5 \times 10^4$ The population was 5×10^4.

Challenge

1. $100 \div 2^4 = 100 \div 16 = 6.25$
   It is 6.25 m from the end of the log.

2. $120 \div 2^3 = 120 \div 8 = 15$
   There were 15 bacteria in the culture 3 days ago.

Unit 5

1. 20.00; 420.00   2. 150.00; 1150.00
3. 96.00; 696.00   4. 15.00; 265.00
5. Amount $= 1000 + 1000 \times \dfrac{1.5}{100} \times 1.5 = 1022.50$
   Peggy received $1022.50.
6. Amount $= 2000 + 2000 \times \dfrac{7}{100} \times 5 = 2700$
   She can accumulate $2700.00.

7.

	Interest	Principal	Year	Rate
Uno	$68.50	$1000.00	1	6.85% ← better
Duo	$6.50	$100.00	1	6.5%

Bank Uno offers the better rate.

8.

Interest	Year	Interest per year	Principal
$4000.00	5	$800.00	$8000.00

Interest rate : $800 \div 8000 \times 100\% = 10\%$
The simple interest rate paid was 10%.

9.

Principal	Rate	Interest per year
$1500.00	10%	$150.00

Time (year) : $300 \div 150 = 2$
She paid back the loan after 2 years.

10.

	Interest	Year	Principal	Interest per year
Alta	$1000.00	2	$2000.00	$500.00
Bee	$1400.00	3	$2000.00	$466.67

Bill should choose Bank Bee.

Challenge

Assume the principal = $1000.00
Rate = $(2000 - 1000) \div 1000 \div 12 \times 100\% = 8.3\%$
The simple interest rate is 8.3%.

Unit 6

1. Length : $170 \times 30\% = 51$ He was 51 cm long at birth.
2. Old allowance : $22 \div (1 + 10\%) = 20$ His old allowance was $20.00.
3. Pay : $3200 \times \dfrac{100 - 35}{100} = 3200 \times \dfrac{65}{100} = 2080$
   One must pay $2080.00.
4. a. No. of card: $64 \times \dfrac{25}{100} + 20 = 36$ John has 36 cards now.
   b. % increase: $\dfrac{36 - 20}{20} \times 100\% = \dfrac{16}{20} \times 100\% = 80\%$
      John's cards has increased by 80%.

5. Alison's score : $\dfrac{12}{15} \times 100\% = 80\%$
   Ruth's score : $\dfrac{15}{18} \times 100\% = 83.3\%$
   Ruth has a higher score.

6. a. She pays : $59.99 + 34.99 \times 2 + 79.99 = 209.96$
      She pays $209.96 altogether.
   b. Tax : $209.96 \times \dfrac{15}{100} = 31.49$
      She pays $31.49 for tax.
   c. Total : $209.96 + 31.49 = 241.45$
      The total cost is $241.45.

7. Price before tax : $9 \div 15\% = 60$
   She paid $60.00 before tax.

8. Price of the skirt : $58.50 \div (1 + 17\%) = 50$
   The price of the skirt before tax is $50.00.

9. Commission : $(225000 + 170000) \times \dfrac{2.5}{100} = 9875$
   Her commission is $9875.00.

10. a. Bird species endangered : $\dfrac{1}{40} \times 100\% = 2.5\%$
       2.5% of the bird species are endangered.
    b. No. of species of birds : $1000 \times 2.5\% = 25$
       25 species of birds in Canada are likely to be endangered.

11. No. of parks : $34 \div 40\% = 85$
    There are 85 national parks in North America.

12. Total money earned : $(2000 + 2000 \times \dfrac{5}{100}) \times (1 + 5\%) = 2205$
    She should expect to earn $2205.00 a year from now.

13. Amount of calories : $210 \div (1 - 30\%) = 300$
    There are 300 calories in a piece of regular cheese of the same size.

14. Percentage Discount : $\dfrac{120 - 102}{120} \times 100\% = 15\%$
    There will be a 15% discount at Christmas.

15. a. Quarters make up the largest contribution.
    b. Money in dimes : $3400000 \times \dfrac{21}{100} = 714000$
       $714 000.00 are collected in dimes.
    c. The percentages in the 1st row are based on the total value of the revenue, but the percentages in the 2nd row are based on the total number of coins collected.

Challenge

Her account had : $1000 \times \dfrac{108.8}{100} \times \dfrac{104.5}{100} \times \dfrac{102.2}{100} \times \dfrac{100.8}{100}$
$= 1171.27$
Judy had $1171.27 in her account on Dec 31, 1993.

Midway Review

1. A   2. D   3. C   4. A   5. B   6. C

7. Volume : $1000 \times 80 \times 10 = 800\,000$
   The volume of the brick wall is 800 000 cm^3.
8. No. of bricks : $800\,000 \div (20 \times 8 \times 10) = 500$
   500 bricks are needed.
9. Total weight : $500 \times 2 + 20 \times 25 = 1500$
   The total weight of the wall is 1500 kg.
10. Total cost : $20 \times 500 \div 100 + 15 \times 20 + 30 \times 4 = 520$
    The total cost of the wall is $520.00.
11. Cost of painting :
    $\dfrac{(1000 \times 80 \times 2 + 80 \times 10 \times 2 + 1000 \times 10)}{60000} \times 50 = 143$
    The cost of painting the wall is $143.00.
12. No. of bricks needed : $(1000 \times 160 \times 20) \div (20 \times 8 \times 10) - 500$
    $= 1500$
    1500 more bricks will be needed.
13. Average speed : $(450 + 150) \div 8 = 75$
    His average speed is 75 km/h.
14. Time : $450 \div 75 = 6$ He will reach Capetown at 6:00 p.m.

ISBN: 978-1-897164-21-1

15. Average speed : $(450 + 150) \div 6 = 100$
His average speed is 100 km/h.
16. Time : $(150 + 50) \div 100 = 2$
He will be 50 km beyond Capetown by 2 p.m.
17. Distance apart: $(100 - 75) \times 3 = 75$
They will be 75 km apart after 3 hours of travelling.
18. Distance travelled by Dave : $75 \times 3 = 225$
Distance travelled by Eric : $100 \times 3 = 300$ $\bigg\rbrack 225 + 300 = 525$
No. They won't meet after 3 hours of travelling.
19. Distance travelled by Jim : $30 \times (12 - 7) = 30 \times 5 = 150$
Yes. He will reach Capetown by noon.
20. % of Canadians living in Ontario : $\dfrac{11413600}{30301200} \times 100\% = 37.7\%$
37.7% of Canadians live in Ontario.
21. % of Canadians who are female : $\dfrac{15302300}{30301200} \times 100\% = 50.5\%$
50.5% of Canadians are female.
22. % of population under 14 :
Quebec : 18.5%   Manitoba : 21.4%
Ontario : 20.0%   Saskatchewan : 22.2%
Saskatchewan has the youngest population.
23. 3.0×10^7
24. % of 0-14 males :   % of 15-64 males :   % of 65+ males :
$\dfrac{3064100}{5975800} \times 100\%$ $\dfrac{10345600}{20588300} \times 100\%$ $\dfrac{1589200}{3737100} \times 100\%$
$= 51.28\%$ $= 50.25\%$ $= 42.52\%$
No. The percentage drops. Probably male Canadians die younger.

Unit 7

1. a. $\dfrac{3}{8} + \dfrac{1}{4} = \dfrac{5}{8}$

   b. $\dfrac{3}{5} + \dfrac{1}{3} = \dfrac{14}{15}$

   c. $\dfrac{1}{4} - \dfrac{1}{6} = \dfrac{1}{12}$

2. a. Time : $2\dfrac{1}{2} + 1\dfrac{1}{3} = 3\dfrac{5}{6}$ He spent $3\dfrac{5}{6}$ hours.

   b. Time : $3\dfrac{5}{6} - 2\dfrac{3}{4} = 1\dfrac{1}{12}$ He spent $1\dfrac{1}{12}$ hours less.

3. Pay : $7.95 \times \dfrac{5}{12} = 3.31$ I would pay $3.31.

4. She ate : $240 \times \dfrac{3}{4} + 210 \times \dfrac{2}{3} = 320$
She ate 320 g of chocolate.

5. Apples weigh : $1500 \times \dfrac{2}{3} = 1000$
Assuming all apples weigh the same. $\dfrac{2}{3}$ bag of apples would weigh 1000 g.

6. Bill ate : $\dfrac{3}{14} = \dfrac{24}{112}$ Carol ate : $\dfrac{5}{16} = \dfrac{35}{112}$
Carol ate more.

7. Time : $3 \div \dfrac{1}{2} \times \dfrac{3}{4} = 4\dfrac{1}{2}$
It takes $4\dfrac{1}{2}$ minutes.

8. Adam : $\dfrac{10}{14} = \dfrac{80}{112}$ Bert : $\dfrac{12}{16} = \dfrac{84}{112}$
Bert gets the better grade.

9. Fraction : $1 - \dfrac{1}{3} - \dfrac{1}{4} - \dfrac{1}{5} = \dfrac{13}{60}$
$\dfrac{13}{60}$ of the cars are neither white, black nor red.

10. No. of planks : $12 \div 1\dfrac{1}{4} = 12 \times \dfrac{4}{5} = 9.6$
Janet must buy 10 planks.

11. No. of dresses : $13\dfrac{1}{2} \div 3 = \dfrac{27}{2} \times \dfrac{1}{3} = 4\dfrac{1}{2}$
She can make 4 dresses.

12. No. of people : $10 \div (\dfrac{1}{2} - \dfrac{1}{3}) = 10 \div \dfrac{1}{6} = 60$
The room holds 60 people when full.

13. No. of textbooks : $90 \div 4\dfrac{1}{4} = 90 \times \dfrac{4}{17} = 21R3$
21 textbooks will fit the shelf.
14. Average : $(\dfrac{1}{4} + \dfrac{3}{4}) \div 2 = \dfrac{1}{2}$
The average of $\dfrac{1}{4}$ and $\dfrac{3}{4}$ is $\dfrac{1}{2}$.
15. a. 70, 93, 140, 175, 210
   b. No. of servings : $350 \div 140 = 2\dfrac{1}{2}$ $2\dfrac{1}{2}$ servings.
16. a.
   $1970 \xrightarrow{+\frac{17}{20}} 1980 \xrightarrow{-\frac{7}{20}} 1990 \xrightarrow{-\frac{1}{5}} 2000$
   The fastest population growth was from 1970 to 1980.
   b. No. of people : $(2\dfrac{17}{20} - 2) \times 10000 = 8500$
   There were 8500 more people.
   c. Percentage change : $\dfrac{(2\dfrac{3}{10} - 2\dfrac{17}{20}) \times 10000}{2\dfrac{17}{20} \times 10000} \times 100\% = -19.3\%$
   The percentage change of population is -19.3%.
   The population decreases because the percentage change of population is a negative figure.
17. a. No. of students : $66 \div \dfrac{2}{17} = 561$
   There are 561 students.
   b. Fraction of girls : $\dfrac{30 \times (1 - \dfrac{1}{2}) + 36 \times (1 - \dfrac{2}{3})}{30 + 36}$
   $= \dfrac{27}{66} = \dfrac{9}{22}$
   $\dfrac{9}{22}$ of the students are girls.
   c. No. of boys: $30 \times \dfrac{1}{2} + 36 \times \dfrac{2}{3} = 39$
   There are 39 boys.
   d. % of students who are boys : $\dfrac{39}{66} \times 100\% = 59.1\%$
   59.1% of the Grade 7 students are boys.
   e. Time : $(\dfrac{1}{5} + \dfrac{1}{4} + \dfrac{1}{3} + \dfrac{1}{4}) = 1\dfrac{1}{30}$
   A Grade 7 student spends $1\dfrac{1}{30}$ h on homework each day.
   f. Time : $1\dfrac{1}{30} \times 5 = 5\dfrac{1}{6}$
   A Grade 7 student spends $5\dfrac{1}{6}$ h on homework during weekdays.
   g. % of students : $\dfrac{20 + 15}{66} \times 100\% = 53.0\%$
   53.0% of the students scored 80 marks or over.
18. a. No. of pages : $320 - 320 \times \dfrac{1}{4} - 320 \times \dfrac{1}{8} = 200$
   He still has 200 pages to read.
   b. No. of pages : $320 - 320 \times \dfrac{1}{4} - (320 - 320 \times \dfrac{1}{4}) \times \dfrac{1}{8} = 210$
   She still has 210 pages to read.
   c. Andrew reads 40 pages while Anna reads 30 pages on the second day. Andrew reads 10 pages more than Anna does.
19. a. Change : $32\dfrac{3}{4} - 37\dfrac{1}{2} = -4\dfrac{3}{4}$
   There is a decrease of $4\dfrac{3}{4}$.
   b. Stock price : $32\dfrac{3}{4} - 4\dfrac{3}{4} = 28$
   The stock price would be $28.00 on Wednesday.

Challenge

Stock price : $1\dfrac{3}{8} - \dfrac{5}{8} + \dfrac{1}{4} - \dfrac{7}{8} = \dfrac{1}{8}$
My stock is worth $ $\dfrac{1}{8}$ more than at first.

Unit 8

1. John got : $150 \div 0.68 = 220.59$ John got C$220.59.
2. Jim got : $78 \times 0.68 = 53.04$ Jim got US$53.04.
3. a. 69.99   b. 14.99   c. 4.99 for 12 cans
   d. 25.99 + 5% tax

4. No. of comic books : 12.50 ÷ 2.54 ≈ 4.92
   I can buy 4 comic books.
5. Weekly expenditure : (320 ÷ 100) x 7 x 0.63 ≈ 14.11
   He spent about $14.11 on gas weekly.
6. Money left : 100 – (100 – 15.49 x 5) = 22.55
   He had $22.55 left for lunch.
7. a. He spent : (16.99 x 3 + 24.25 x 2) x 1.15
              = (50.97 + 48.5) x 1.15 = 114.39
      He spent $114.39.
   b. Change : 150 – 114.39 = 35.61
      He got $35.61 change.
8. Each would get : 4.75 ÷ 8 ≈ 0.59
   Each would get about $0.59.
9. No. of nickels : 3.48 ÷ 0.05 ≈ 69
   The maximum number of nickels she has is 69.
10. Distance apart : 3.73 x 150 = 559.5
    The towns are 559.5 km apart on land.
11. Length of a Mars year : 1.88 x 365.3 ≈ 686.76
    The length of a Mars year is 686.76 days.
12. a. Time : 70.83 ÷ 48 ≈ 1.48
       It took about 1.48 hours.
    b. No. of orbits : (10 ÷ 70.83) x 48 ≈ 6.78
       She made 6.78 orbits in 10 hours.
13. $4.5 ÷ (1 + \frac{1}{4}) = 3.6$
    The number is 3.6.
14. 0.357; 0.308; 0.333; 0.313
15. New average : 5 ÷ (14 + 3) = 0.294
    His new batting average was 0.294.
16. No. of hits : 0.375 x (13 + 3) – 4 = 6 – 4 = 2
    He had 2 more hits.
17. No. of times at bat : (4 + 1) ÷ 0.333 – 12 ≈ 3
    He had 3 times at bat that day.
18. Batting average : (0.357 + 0.308 + 0.333 + 0.313) ÷ 4 ≈ 0.328
    The batting average of the four batters was about 0.328.
19. Time : (35.2 + 27.4 + 48.7) ÷ 65.5 = 1.70
    He takes 1.70 h (1 h 42 min) to reach Greenpark.
20. Average speed : (12.5 + 28.3 + 27.4 + 48.7) ÷ 2.2 = 53.136
    His average speed is 53.136 km/h.
21. Distance : 2.90 x 70 ÷ 2 = 101.5
    From Huntsville to Greenpark via Brownsville :
    52.8 + 48.7 = 101.5
    He took the route via Brownsville.

Challenge

Each group has 1 loonie and 2 dimes; its value is $1.20.
No. of groups of coins : (15.35 – 19 x 0.05) ÷ 1.20 = 12
There are 12 loonies and 24 dimes.

Unit 9

1. always
2. sometimes
3. never
4. never
5. always
6.

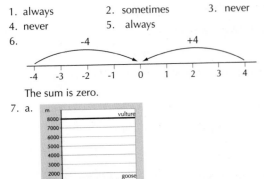

The sum is zero.

7. a.

b. Sum : (+8000) + (+1000) + (-1200) = +7800
   The sum is +7800.
c. Difference : +8000 – (+1000) = +7000
   The difference is +7000 m.
d. Difference : +8000 – (-1200) = +9200
   The difference is +9200 m.
8. a. Mt Logan is +5900 m; the trench is -11 000 m.
   b. Distance apart : -11000 + 5900 = -5100
      No. Mt Logan would still be 5100 m below sea level.
9. a. Changes : $+23\frac{1}{8} – 2\frac{1}{4} + 2\frac{3}{4} – 3\frac{1}{2} + 2 – 3\frac{3}{4}$
   b. Money change : $(-2\frac{1}{4} + 2\frac{3}{4} – 3\frac{1}{2} + 2 – 3\frac{3}{4}) x 1000$
                  $= -4\frac{3}{4} x 1000 = -4750$
      He lost $4750.00.
10. a. Mercury is the coldest.
    b. Difference : -90 – (-123) = 33
       The difference is 33°C.
    c. -90°C > -123°C > -184°C or
       -184°C < -123°C < -90°C
11. a. Temperature : -35 – (-19) = -16
       The temperature in Calgary was -16°C.
    b. Difference : -19 – (-16) = -3
       The difference was 3°C.
12. a. Jan 1 : -5   Jan 7 : -4
       The change is -4 – (-5).
    b. Temperature : -4 + (-7) = -11
       The temperature on Jan 8 was -11°C.
13. Football game : +6 + (-3) + (-2) + (-4) = -3
    He had lost 3 m overall.
14. a. Mr Smith's score :
       0 + (+2) + (-1) + (-2) + (0) + (+1) + (+1) + (-1) + (+2)
       He scored 2 over par.
    b. His score : +36 + (+2) = +38
       His score was +38.

Challenge

1. The sum of the integers is negative, so the integers are negative.
   (-3, -6); -3 – (-6) = 3 ✘
   (-2, -7); -2 – (-7) = 5 ✔
   The integers are -2 and -7.
2. The average of the numbers is 9; the sum of the numbers is 18.
   (12, 6); (12 + 6) ÷ 2 = 9, 12 – 6 = 6 ✘
   (13, 5); (13 + 5) ÷ 2 = 9, 13 – 5 = 8 ✔
   The integers are 5 and 13.

Unit 10

1. a.

Distance (km)	5	10	15
Cost ($)	6	10	14

+4   +4

The cost of a 15 km ride is $14.00.

b.

Distance (km)	15	20	25	30	35	40
Cost ($)	14	18	22	26	30	34

The cost of a 40 km ride is $34.00.

c. I can travel 20 km for $18.00.

2.

Week	0	1	2	3	4	5
Weight (g)	100	200	400	800	1600	3200

x2 x2 x2 x2 x2

It takes 5 weeks to reach at least 3 kg.

ISBN: 978-1-897164-21-1

3.

Year	1960	1970	1980	1990	2000	2010	2020
No. of students	500	450	550	500	600	550	650

−50 +100 −50 +100 −50 +100

There will be 650 students in 2020.

4. a.

Week	0	1	2	3	4	5	6	7	8
Value	2	2.5	3	3.5	4	4.5	5	5.5	6

It will be worth $6.00 after 8 weeks.

b.

Week	8	9	10
Value	6	6.5	7

It will be worth $7.00 after 10 weeks.

5. a.

Day	1	2	3	4
Weight (g)	5	9	17	33

+4 +8 +16

The difference of weight doubles every time.

b.

Day	4	5	6	7
Weight (g)	33	65	129	257

It weighs 257 g on the seventh day.

6.

Year	6th	5th	4th	3rd	2nd	1st	0
Value	45	42	39	36	33	30	27

The original price was $27.00.

7. a.

Day	1	2	3	4	5	6	7	8	9	10
No. of books	52	48	50	46	48	44	46	42	44	40

−4 +2 −4 +2 −4 +2 −4 +2 −4

There will be 40 books on the tenth day.

b.

Day	10	11	12	13	14
No. of books	40	42	38	40	36

36 books will remain on the shelf after 14 days.

8. a.

$3 \xrightarrow{\times 2} 6 \xrightarrow{\times 2} 12 \xrightarrow{\times 2} 24 \xrightarrow{\times 2} 48 \xrightarrow{\times 2} 96$

b.

$96 \xrightarrow{\times 2} 192 \xrightarrow{\times 2} 384$

The eighth number is 384.

9. Every number in the series is the sum of the preceding two numbers. The next 5 terms are 21 (8 + 13), 34 (13 + 21), 55 (21 + 34), 89 (34 + 55) and 144 (55 + 89).

10. September 1 could be a Sunday, a Monday or a Thursday.

11. a.

$1 \xrightarrow{+3} 4 \xrightarrow{+5} 9 \xrightarrow{+7} 16$

Differences increase by 2.

$1 (1^2) \longrightarrow 4 (2^2) \longrightarrow 9 (3^2) \longrightarrow 16 (4^2)$

Each number is the square of a whole number.

b.

$1 \xrightarrow{+3} 4 \xrightarrow{+5} 9 \xrightarrow{+7} 16 \xrightarrow{+9} 25 \xrightarrow{+11} 36$

$1 (1^2) \longrightarrow 4 (2^2) \longrightarrow 9 (3^2) \longrightarrow 16 (4^2) \longrightarrow 25 (5^2) \longrightarrow 36 (6^2)$

The sixth number is 36.

12.

$1 \xrightarrow{+2} 3 \xrightarrow{+3} 6 \xrightarrow{+4} 10 \xrightarrow{+5} 15 \xrightarrow{+6} 21 \xrightarrow{+7} 28$

The seventh triangular number is 28.

13. a. 1 + 3 + 5 + 7 + 9 + 11 = 36

The sixth number is 36; there are 6 numbers in the series.

b. 1 + 3 + 5 + 7 + 9 + 11 + 13 + 15 + 17 + 19 = 100

The tenth number is 100.

Challenge

101st triangular number = sum of 100th triangular number and 101
= 5050 + 101 = 5151

The 101st triangular number is 5151.

Unit 11

1. Let n be the score.
$3.2n = 86.08 \longrightarrow n = 26.9$
The score was 26.9.

2. Let n be the snowfall in Vancouver.
$5n - 10 = 360 \longrightarrow 5n = 370 \longrightarrow n = 74$
The snowfall in Vancouver was 74 cm.

3. Let n be the smaller number.
$n + 2 + n = 40 \longrightarrow 2n = 38 \longrightarrow n = 19$
The numbers are 19 and 21.

4. Let n be the time needed.
$25n + 30 = 80 \longrightarrow 25n = 50 \longrightarrow n = 2$
It takes 2 hours.

5. Let n be the length of the third side.
$6 + 13 + n = 28 \longrightarrow n = 9$
The length of the third side is 9 cm.

6. Let n be the distance travelled.
$0.1n + 30 = 50 \longrightarrow 0.1n = 20 \longrightarrow n = 200$
The Kims travelled 200 km.

7. Let n be the distance walked by a doctor daily.
$2n + 1.2 = 6.8 \longrightarrow 2n = 5.6 \longrightarrow n = 2.8$
A doctor walks 2.8 km daily.

8. Let n be the number of exercise classes Anna attends.
$2n + 24 = 40 \longrightarrow 2n = 16 \longrightarrow n = 8$
Anna attends 8 exercise classes.

9. Let n be the number.
$3n - 15 = 21 \longrightarrow 3n = 36 \longrightarrow n = 12$
The number is 12.

10. Let n be the electricity consumed.
$0.2n + 6 = 52 \longrightarrow 0.2n = 46 \longrightarrow n = 230$
Mr Mark consumed 230 kWH of electricity.

11. Let n be the time needed to recover the bulb cost.
$1.5n + 20 = 50 \longrightarrow 1.5n = 30 \longrightarrow n = 20$
It takes 20 months for them to recover the bulb cost.

12. Let n be the number of girls.
$n + 2n - 31 = 83 \longrightarrow 3n = 114 \longrightarrow n = 38$
No. of boys = 83 − 38 = 45
There are 45 boys and 38 girls.

13. Let n be the original number.
$(n - 15) \times 2 = 22 \longrightarrow n - 15 = 11 \longrightarrow n = 26$
The number is 26.

14. a. Let n be the cost of each tea bag.
$24n + 0.5 = 1.7 \longrightarrow 24n = 1.2 \longrightarrow n = 0.05$
Each tea bag costs $0.05.
b. Cost : 75 × 0.05 + 0.5 = 4.25
The cost is $4.25.

15. Let n be the maximum allowable wattage of each surrounding bulb.
$120 + 8n = 600 \longrightarrow 8n = 480 \longrightarrow n = 60$
The maximum allowable wattage is 60 W.

16. Let n be the time taken.
$-80 + 2n = -14 \longrightarrow 2n = 66 \longrightarrow n = 33$
He would take 33 seconds to reach a depth of 14 m.

17. Let n be the number of pennies saved by James.
$1.2n + 520 = 1636 \longrightarrow 1.2n = 1116 \longrightarrow n = 930$
James has saved 930 pennies.

18. Let n be the number of hours.
$6.8n + 20 = 802 \longrightarrow 6.8n = 782 \longrightarrow n = 115$
Cindy worked 115 hours.

ISBN: 978-1-897164-21-1

Challenge

Let n be the number of months that the Wong family have to take to recover the cost.

$120n - 120 \times \frac{2}{3} n = 600 \longrightarrow 120n - 80n = 600 \longrightarrow$

$40n = 600 \longrightarrow n = 15$

The Wong family have to take 15 months to recover the cost.

Final Review

1. Fraction of the students who are boys : $\frac{40}{70} = \frac{4}{7}$

   $\frac{4}{7}$ of the Grade 7 students are boys.

2. Fraction of the students who are 13 years old :

   $\frac{70 - 56}{70} = \frac{14}{70} = \frac{1}{5}$

   $\frac{1}{5}$ of the Grade 7 students are 13 years old.

3. No. of students : $40 \times \frac{1}{8} + 30 \times \frac{1}{3} = 5 + 10 = 15$

   15 Grade 7 students have joined the Environmental Club.

4. Fraction of students who joined the club : $\frac{15}{70} = \frac{3}{14}$

   $\frac{3}{14}$ of the Grade 7 students who have joined the Environmental Club.

5. Fraction of female members : $\frac{10}{15} = \frac{2}{3}$

   $\frac{2}{3}$ of the club members are female.

6. There is a bigger fraction of girls in the club than in class. It shows that more girls are concerned about the environment than boys.

7. Money collected from girls : $19.5 - 1.5 \times 5 = 12$

   No. of girls who paid : $12 \div 1.5 = 8$

   8 female members have paid.

8. 15, 18; 20, 21; 25, 24

9. a. $0 \longrightarrow 5 \longrightarrow 10 \longrightarrow 15 \longrightarrow 20 \longrightarrow 25$

   The number of candles sold by girls is 5x.

   b. $9 \longrightarrow 12 \longrightarrow 15 \longrightarrow 18 \longrightarrow 21 \longrightarrow 24$

   The number of candles sold by boys is 9 + 3x.

   c. $9 \longrightarrow 17 \longrightarrow 25 \longrightarrow 33 \longrightarrow 41 \longrightarrow 49$

   The number of candles sold by the club members is 9 + 8x.

10. No. of candles : $9 + 8 \times 10 = 9 + 80 = 89$

    89 candles will be sold altogether.

11. Money raised : $89 \times 0.8 = 71.2$

    $71.20 would be raised in 10 days.

12. a. No. of recyclable cans : $240 \times 1.2 \times \frac{3}{4} = 216$

    216 recyclable cans will be collected each week.

    b. Money raised : $216 \div 100 \times 6 = 12.96$

    The club will raise $12.96 in a week.

13. Each member paid : $(16.99 + 16.99 \times 15\%) \times 2 \div 12 = 3.26$

    Each member paid $3.26.

14. Ottawa is the hottest city.

15. Inuvik is the coldest city.

16. Yellowknife has the most extreme weather conditions.

17. Edmonton, Inuvik, Winnipeg and Yellowknife have an annual range of temperature over 38°C.

18. Range : $20 - (-1) = 21$

    The annual range of temperature of Halifax is 21°C.

19. Average : $(-7 + 8 + 22 + 10) \div 4 = 8.25$

    The average temperature of Banff is 8.25°C.

20. Average : $(-25 - 1 + 21 + 1) \div 4 = -1$

    Yellowknife has an average temperature that is below freezing.

21. Change : $-5 - (-8) = +3$

    The temperature rises by 3°C.

22. a. Average :

    $(-7 - 6 - 17 - 12 - 1 - 26 - 6 - 16 - 14 - 25) \div 10 = -13$

    The average temperature of the 10 cities in January is -13°C.

b. Average :

$(10 + 12 + 11 + 7 + 14 - 5 + 13 + 4 + 12 + 1) \div 10 = 7.9$

The average temperature of the 10 cities in October is 7.9°C.

23. C	24. B	25. D
26. A	27. B	28. C
29. B	30. A	31. C

ISBN: 978-1-897164-21-1

Unit 1

1. Area : $(3 \times 1.4 - 1.4 \times 1.4) \div 2 = 2.24 \div 2 = 1.12$
   The area is 1.12 m^2.
2. Area : $50 \times 4 \times 50 \times 2 = 20000$ The area is 20 000 cm^2.
3. The perimeter of the base : $(60 + 36) \times 2 = 192$
   No of times : $3 \times 1000 \div 192 = 15.625$
   They must walk around 16 times.
4. a. Possible dimensions : $16 = 1 \times 16; 2 \times 8; 4 \times 4$
      Possible perimeters : $(1 + 16) \times 2 ; (2 + 8) \times 2; (4 + 4) \times 2$
      The possible lengths are 34 m, 20 m and 16 m.
   b. The minimum length of fence is 16 m.
5. a. Sum of width and length : $20 \div 2 = 10$
      Possible widths and lengths : $(5,5); (6,4); (7,3); (8,2); (9,1)$
      Possible areas : $(5 \times 5); (6 \times 4); (7 \times 3); (8 \times 2); (9 \times 1)$
      The possible values for the areas are 25 m^2, 24 m^2, 21 m^2, 16 m^2 and 9 m^2.
   b. Difference : $25 - 9 = 16$ The difference is 16 m^2.
   c. • Rectangles with the same perimeter may have different areas.
      • The greater the difference between the length and width of a rectangle, the smaller the area of the rectangle.
6. No. of tiles : $100 \times 100 \div 1 \times 1 = 10000$ He needs 10 000 tiles.
7. Perimeter : $10 + (16 - 8) + 1 + 8 + (10 - 1 - 7) + 16 = 45$
   Cost : $45 \times 15 = 675$ The fencing costs $675.00.
8. Area : $(5 + 1 + 1) \times (7 + 1 + 1) - 5 \times 7 = 28$ The area is 28 m^2.
9. Area of backyard : $1 \times (16 - 8) + (10 - 1) \times 16 = 8 + 144 = 152$
   Area to be planted : $152 - (5 + 1 + 1) \times (7 + 1 + 1) = 89$
   Cost : $89 \times 12 = 1068$ The cost of the grass will be $1068.00.
10. Floor area : $4 \times 5 = 20$
    Cost : $20 \times 70 = 1400$ The carpet costs $1400.00.
11. Total area : $(5 \times 2.5) \times 2 + (4 \times 2.5) \times 2 - 7 + 5 \times 4 = 58$
    Amount of paint needed : $58 \div 5 = 11.6$
    Cost of paint : $11.6 \times 8.99 = 104.28$ The cost of the paint is $104.28.
12. Total area : $10 \times 2 \times 8 = 160$ The total area is 160 cm^2.
13. Perimeter : $24 \div 3 \times 4 = 32$ The perimeter is 32 cm.
14. Average area : $9 \times 4 \div 3 = 12$ The average area is 12 cm^2.
15. a. Perimeter : $12 + 12 + 6 = 30$ The perimeter is 30 cm.
    b. Area : $6 \times 11.62 \div 2 = 34.86$ The area is 34.86 cm^2.
16. a. Perimeter : $50 \times 6 = 300$ The perimeter is 300 cm.
    b. Area : $88 \times 24 \div 2 \times 2 + 50 \times 88 = 6512$ The area is 6512 cm^2.
17. Shaded area : $2 \times 3 \div 2 = 3$ The area is 3 cm^2.
18. Area of square = area of triangle : $5 \times 5 = 25$
    Height of triangle : $25 \times 2 \div 10 = 5$ The height of the triangle is 5 cm.
19. Area : $3 \times 4 \div 2 + 12 \times 5 \div 2 = 6 + 30 = 36$
    Perimeter : $4 + 3 + 13 + 12 = 32$
    The area is 36 cm^2 and the perimeter is 32 cm.
20. Area : $6 \times 5 + 6 \times 3 \div 2 = 39$ The area is 39 m^2.
21. Total area : $(30 + 36) \times 30 \div 2 \times 2 = 1980$
    Their total area is 1980 cm^2.
22. Total area : $(44 + 48) \times 30 \div 2 \times 2 = 2760$
    Their total area is 2760 cm^2.
23. Total area : $(5 + 10) \times 6 \div 2 \div 12 \times 3 = 11.25$
    The total area is 11.25 cm^2.
24. Shaded area : $(1.2 + 2) \times 1 \div 2 \times 2 = 3.2$ The area is 3.2 cm^2.
25. Area : $(0.5 + 0.7) \times (1 - 0.7) \div 2 + 0.7 \times 0.7 = 0.67$
    Its cross-sectional area is 0.67 m^2.

Challenge

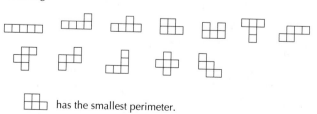

has the smallest perimeter.

Unit 2

1. $1 m^3 = 100$ cm x 100 cm x 100 cm $= 1000000$ cm^3
   There are 1 000 000 cm^3 in 1 m^3.
2. $1 m^2 = 100$ cm x 100 cm $= 10000$ cm^2
   There are 10000 cm^2 in 1 m^2.
3. Volume : $12 \times 12 \times 12 = 1728$;   Capacity : $1728 \div 1000 = 1.728$
   It can contain 1.728 L of water.
4. Area of each surface : $54 \div 6 = 9$
   Length of each side : $\sqrt{9} = \sqrt{3 \times 3} = 3$
   The dimensions are 3 cm x 3 cm x 3 cm.
5.
6. Before k is removed, there are $4 \times 6 = 24$ surfaces.
   After k is removed, there are still 24 surfaces.
   No effect. The total number of surfaces remains the same.
7. a. The total surface area will increase.
   b. 4 more surfaces will be exposed. Total area : $2 \times 2 \times 4 = 16$
      There is an increase of 16 cm^2.
8. a. Dimensions : $(12 \div 4)$ by $(12 \div 4)$ by $(12 - 3 \times 2)$
      The dimensions are 3 cm by 3 cm by 6 cm.
   b. Total surface area : $3 \times 3 \times 2 + 3 \times 6 \times 4 = 18 + 72 = 90$
      The total surface area is 90 cm^2.
   c. Volume : $3 \times 3 \times 6 = 54$ The volume is 54 cm^3.
9. Total surface area of the gift :
   $20 \times 30 \times 2 + 8 \times 30 \times 2 + 20 \times 8 \times 2 = 2000$
   Area of the paper : $100 \times 100 = 10000$ $(10000 > 2000)$
   She will have enough paper.
10. a. Assume each side of the cube is 1 cm.   Total surface area : $1 \times 6 = 6$
       Total surface area of a 2 cm cube : $2 \times 2 \times 6 = 24$
       No. of times : $24 \div 6 = 4$ The surface area has increased by 4 times.
    b. Volume : $1 \times 1 \times 1 = 1$;   New volume : $2 \times 2 \times 2 = 8$
       No. of times : $8 \div 1 = 8$ The volume has increased by 8 times.
11. A cube has 6 faces. Surface area of each face : $216 \div 6 = 36$
    Length of each side : $\sqrt{36} = 6$;   Volume : $6 \times 6 \times 6 = 216$
    The volume is 216 m^3.
12. a. Amount of wood : $1.5 \times 1 \times 2 + 1.2 \times 1 \times 2 + 1.5 \times 1.2 \times 2 = 9$
       The amount of wood required is 9 m^2.
    b. Volume : $1.2 \times 1.5 \times 1 = 1.8$ The volume is 1.8 m^3.
13. Total area to be painted : $5 \times 4 \times 1 + 4 \times 3 \times 2 + 5 \times 3 \times 2 - 3 = 71$
    Number of cans needed : $71 \times 2 \div 36 = 3.94$
    I need 4 cans of 4 L paint.
14. No. of dice : $(10 \times 10 \times 10) \div (2 \times 2 \times 2) = 125$
    125 dice can be placed in the box.
15. Volume : $30 \times 16 \times 23 \times 3 + 20 \times 16 \times 23 \times 1.5 = 44160$
    The minimum volume of the container is 44 160 cm^3.
16. Box A : $11.99 \div (17 \times 30 \times 30) = 0.00078$
    Box B : $6.99 \div (15 \times 25 \times 25) = 0.00075$ $(0.00075 < 0.00078)$
    Box B is a better buy.
17. No. of servings : $(31 \times 20 \times 7) \div 175 = 24R140$
    It contains 24 servings.
18. 27m^3/h for 1 workman; 54 m^3/h for 2 workmen.
    Time taken : $(6 \times 6 \times 6) \div 54 = 4$ They would take 4 hours.
19. Volume of the pyramid : $5 \times 5 \times 12 \times \frac{1}{3} = 100$
    New water level : $12 + 100 \div (5 \times 5) = 12 + 4 = 16$
    The new water level will be 16 cm.
20. There are 15 rectangular prisms with $(30 \times 20 \times 150)$cm^3.
    Volume of cement needed : $30 \times 20 \times 150 \times 15 = 1350000$
    The volume of cement needed is 1 350 000 cm^3 (1.35m^3).
21. Length of each side of cardboard : $\sqrt{25} = 5$
    Capacity of the open box : $(5 - 2) \times 1 \times (5 - 2) = 9$
    The capacity is 9 mL.

ISBN: 978-1-897164-21-1

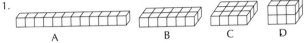

22. Volume of the tank : 3 x 1 x 1 – 1 x 0.5 x 1 = 2.5
    The volume of the tank is 2.5 m^3 (2 500 000 cm^3).
    Time needed : 2500000 ÷ (20 x 1000) = 125
    It would take 125 minutes (2 hr 5 min).
23. No. of cubes : 4 x 4 x 2 + 4 x 2 x 4 = 64
    64 cubes will have just one face painted.

Challenge

1.

    A   B   C   D

2. A : 1 x 12 x 4 + 1 x 1 x 2 = 50
    B : 2 x 6 x 2 + 1 x 6 x 2 + 1 x 2 x 2 = 40
    C : 3 x 4 x 2 + 1 x 4 x 2 + 1 x 3 x 2 = 38
    D : 2 x 3 x 4 + 2 x 2 x 2 = 32
    The surface area of each prism is 50 cm^2, 40 cm^2, 38 cm^2, 32 cm^2.
3. Yes. The volume of each prism is 12 cm^3.

Unit 3

1. False
2. True
3. False
4. True
5. True
6. True
7. The scale is 1:300; 15 m = 1500 cm; 18 m = 1800 cm.
    Dimensions on the blueprint : $\frac{1500}{300} = 5$; $\frac{1800}{300} = 6$
    The dimensions are 5 cm by 6 cm.
8. Ratio of the widths : $\frac{3.6}{3} = \frac{1.2}{1}$
    Ratio of the lengths : $\frac{8.4}{7} = \frac{1.2}{1}$
    Yes. They are similar to each other because corresponding sides are in the same ratio.
9. Ratio of the widths : $\frac{37}{26}$
    Ratio of the lengths : $\frac{27}{21} = \frac{9}{7}$ ($\frac{37}{26} \neq \frac{27}{21}$)
    No. They are not similar to each other because corresponding sides are not in the same ratio.
10. 540 m = 54000 cm
    54 000 is represented by 6 or 9000 is represented by 1.
    The scale is 1:9000.
11. Yes. Both triangles are equal in size and shape.
12. Length : $\frac{3}{2}$ x 1.2 = 1.8; $\frac{3}{1.2}$ x 2 = 5
    The length of the other 2 sides is either 1.8 cm or 5 cm.
13. 10.8:7.2 = X:5; X = $\frac{10.8}{7.2}$ x 5 = 7.5 X is 7.5 cm long.
14. 5.4:7.2 = X:4.2; X = 4.2 x $\frac{5.4}{7.2}$ = 3.15
    A = 120°(corresponding angles) X is 3.15 cm long; angle A is 120°.
15. 9.6:20 = width:14; width = $\frac{9.6}{20}$ x 14 = 6.72
    The width is 6.72 cm.
16. Ratio of the corresponding sides : $\frac{2.4}{3} = \frac{0.8}{1} = \frac{4}{5} = \frac{1.2}{1.5}$
    Yes. They are similar to each other because all sides are proportional.
17. Ratio of the corresponding sides : $\frac{9}{6} = \frac{6}{4}$
    Yes. They are similar to each other because corresponding sides are proportional.
18. No. The two triangles are not necessarily congruent since the sides may be different in length.
19. Yes. The equilateral triangles are similar since corresponding sides are in the same ratio.
20. The area of each shape is 27 square units.

21. △CBA, △CDE, △CFG, △CHI and △CJK are similar triangles.

22. a. Each face is an equilateral triangle.
    b. Yes. All the faces are congruent.

Challenge

1. 1 cm ⌐A   5 cm
   2 cm
           B

2.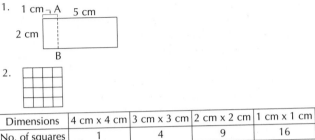

Dimensions	4 cm x 4 cm	3 cm x 3 cm	2 cm x 2 cm	1 cm x 1 cm
No. of squares	1	4	9	16

I can construct 30 squares. 30 squares are similar.
There are 3 congruent groups.

Unit 4

1. Diagrams A and D illustrate a translation.
2. Diagrams B and C illustrate a reflection.
3. Diagrams B and D illustrate a rotation.
4. a. Translation   b. Reflection   c. Rotation
5. Suggested answer:
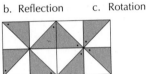

   It is a rotation.
6. It is a rotation.
7. a. A, D; C, E; F, G can match.
    b. B does not have a match.

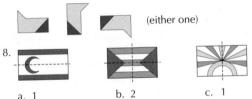
(either one)

8.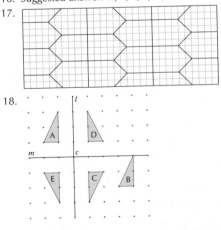

    a. 1   b. 2   c. 1
9. a. B and C have rotational symmetry.
    b. For figures B and C, a $\frac{1}{2}$ turn will rotate the object onto itself.
10. It has rotational symmetry.
11. It is 5.
12. D, A, B, C, E.
13.

14. a. E,*,κ,Φ have a horizontal line of symmetry.
    b. *,п,Φ,m have a vertical line of symmetry.
    c. *,Φ have rotational symmetry.
15. A and E can.
16. Suggested answer: H, L, E, V, N, Z
17.

18.

ISBN: 978-1-897164-21-1

19. a. The parallelograms all have an area of 4 square units.
    b. K to L ⟶ translated 4 right 1 down.
       K to M ⟶ rotated 180° about point *x*.
       K to N ⟶ reflected in the line *l*.

20.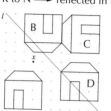

Challenge

1. Suggested answer:

   Yes. A figure with only one line of symmetry does not have rotational symmetry, e.g. an isosceles triangle.

2. Suggested answer:

Midway Review

1. a. Dimensions : 12 ÷ 100 by 12 ÷ 100 by 3 ÷ 100
        = 0.12 by 0.12 by 0.03
     The dimensions are 0.12 m by 0.12 m by 0.03 m (12 cm by 12 cm by 3 cm).
   b. Space : 12 x 12 x 3 = 432
     The space inside the model is 432 cm³.
   c. Area : 12 x 12 x 2 + 12 x 3 x 2 + 12 x 3 x 2 = 432
     432 cm² of cardboard is needed.
   d. Scale : 1:100; No. of times bigger : 100 x 100 x 100 = 1000000
     The space occupied by the cottage is 1 000 000 times bigger than that of the model.
   e. Scale of the model : 1:100
     No. of times bigger : 100 x 100 = 10000
     The interior surface area of the cottage is 10 000 times bigger than that of the model.
   f. D

2. a. There are 9 triangles similar to △ABC.
   b. There are 3 triangles congruent to △ABD.
   c. Area : $\frac{8}{2}$ x $\frac{6}{2}$ = 12   The area is 12 cm².
   d. Area : ($\frac{6}{2}$ + 6) x $\frac{8}{2}$ ÷ 2 = 18   The area is 18 cm².
   e. Total surface area : 8.5 x 6 x 2 + (6 x 8 ÷ 2) x 2 = 150
     The total surface area is 150 cm².
   f. Total surface area : (6 x 8 ÷ 2) x 2 + 8.5 x 6 x 2 + 6 x 6 x 5
                = 48 + 102 + 180 = 330
     The total surface area is 330 cm².
   g. 12 2-cm cubes will have 2 painted faces.
   h. Height : 20 – $\frac{6 \times 6 \times 6}{10 \times 8}$ = 17.3
     The original height was 17.3 cm.
   i.

3. a. Area : 6 x 4 = 24   The area is 24 m².
   b. Area : (6 x 4) + (4 x 2.5 x 2) + (6 x 2.5 x 2) – (1 x 2) – (1 x 1) = 71
     The area is 71 m².

c. No. of tiles : 24 x 100 x 100 ÷ (25 x 25) = 384
     384 tiles are needed.
   d. 1 box has 50 tiles; 8 boxes are needed for 384 tiles.
     Cost : 21.95 x 8 = 175.60   The cost is $175.60.
   e. No. of tiles : 8 x 50 – 384 = 16   16 tiles are left over.
   f. Total area to be painted = 71
     Amount of paint : 71 ÷ 4 x 2 = 35.5
     35.5 L of paint is needed.
   g. No. of cans of paint needed : 35.5 ÷ 4 = 8.875 (9 cans)
     Cost : 27.99 x 9 = 251.91
     The cost of the paint is $251.91.
   h. Amount of paint : 9 x 4 – 35.5 = 0.5
     0.5 L of paint will be left over.

4.
   Cut along the dotted lines — Cut into halves

   I would cut the cube diagonally first and then cut it into halves.

5. XY:8 = 2.4:4;   XY = $\frac{2.4 \times 8}{4}$ = 4.8

   ZY:5 = 2.4:4;   ZY = $\frac{2.4 \times 5}{4}$ = 3

   Perimeter of △XYZ : 2.4 + 4.8 + 3 = 10.2
   The perimeter of △XYZ is 10.2 cm.

6. I would not consider tile D because pentagons can't form a tiling pattern.

7.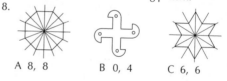

   She could use some square-shaped bricks along with the octagon-shaped bricks to form the tiling pattern.

8.

   A 8, 8   B 0, 4   C 6, 6

Unit 5

1. a. B   b. D   c. A   2. C
3. a. The greatest increase was between 1995 and 1996.
   b. There was a decrease between 1993 and 1994.
   c. No. It was because a separate circle would be needed for each year.
   d. About 33% of female workers would be expected in the year 2002.
4. Graph A : The horizontal axis is not labelled and marked properly. It is meant to convey the trend of sales of ruby red wine.
   Graph B : The vertical scale should start at 0 and the title of the graph is inappropriate. It is meant to convey the number of absent employees in a week.
   Graph C : The scale along the horizontal axis is not consistent. It is meant to convey the trend of prices of 1 litre of gas in Ontario.
5. a. i. I would use graph B because it shows a steady increase in profits made by the company.
     ii. I would use graph C because the line showing the profits made by the drug company rises sharply from 1990 to 1999.
     iii. I would use graph A because the line showing the profits made by the drug company increases slowly from 1990 to 1999.
   b. The profit might be about 900 million dollars in 2000.

ISBN: 978-1-897164-21-1

c. Profits made by the drug company between 1990 and 1999:

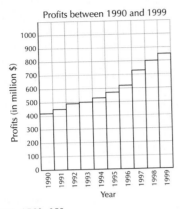

Profits between 1990 and 1999

6. a. 60°, 90°, 150°, 60°

b. Students' Activities

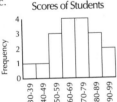

Aerobics | Cycling
Running
Swimming

c. Suggested answer:
Bar graph or Pictograph

d. A line graph would not be suitable.

7. a.

tens	ones
3	7
4	8
5	0, 2, 8
6	4, 6, 7, 8
7	2, 5, 5, 9
8	3, 5, 9
9	1, 5

b.

Score	Frequency
30 - 39	1
40 - 49	1
50 - 59	3
60 - 69	4
70 - 79	4
80 - 89	3
90 - 99	2

c.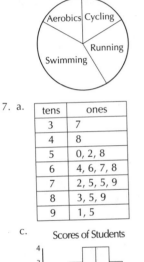

Scores of Students

d. Scores of Students

e. There are 18 students and 16 got 50 or more marks. The passing percentage of the students was $\frac{16}{18}$ x 100% = 88.9%.

Challenge

1. It reached its highest value in October and its lowest value in August.

2. Percentage change : $\frac{280 - 290}{290}$ x 100% = -3.45%

There was a decrease of 3.45%.

3. Percentage change : $\frac{340 - 250}{250}$ x 100% = 36%

There was an increase of 36%.

Unit 6

1. Suggested answers:
   a. 13, 16, 16   b. 15, 15, 21   c. 14, 15, 19
   d. 10, 14, 15   e. 15, 16, 16, 17   f. 6, 7, 8

2. a. Mean : (15 + 20 + 20 + 15 + 18 + 23 + 18 + 24 + 18) ÷ 9 = 19
   The mean is 19.
   b. Median : 15, 15, 18, 18, 18, 20, 20, 23, 24
   The median is 18 and the mode is 18.

c. Mean : (15 + 15 + 18 + 18 + 18 + 20 + 20 + 23 + 24 + 29) ÷ 10
= 20
Median : (18 + 20) ÷ 2 = 19
The mean would increase. The median would be the average of 18 and 20 instead of 18. The mode would not be affected.

3. a. Mean : (7 + 9 + 11 + 12 + 14x2 + 18 x 2 + 19 + 21 + 23 + 24 + 26x2 + 28x2 + 35x3 + 38x2 + 42x2) ÷ 23 = 563 ÷ 23 = 24.48
   The mean is 24.48.
   b. The median is 24. The mode is 35.

4. a. Mean : (30000x12 + 40000x2 + 45000x3 + 60000x2 + 100000) ÷ 20 = 39750   The mean is 39 750.
   The median is 30 000.   The mode is 30 000.
   b. I would use the mean because it is higher than the mode and the median. It would make the salary look more attractive.
   c. It would affect the mean but not the median or mode.
   d. The median or mode best represents the salaries because it is the most likely salary you would earn when you start working there.

5. Average : (1.25x5 + 1.25x3 + 0.95x4) ÷ 12
= (6.25 + 3.75 + 3.8) ÷ 12 = 1.15
The average price they paid was $1.15 per ice cream cone.

6. Mean : (25 + 32x2 + 24x3 + 28x4 + 0) ÷ 11 = 24.82
The mean allowance of the children is $24.82.

7. The mean is 73.2. The median is 72. The range is 78 – 68 = 10.

8. a. The winning heights increased except in 1992.
   b. Height : (1.83 + 1.92 + 1.93 + 1.96 + 2.08 + 2.09 +2.07 + 2.15) ÷ 8
   ≈ 2
   The approximate winning height is 2 m.
   c. If the trend continued, the height might be 2.16 m.
   d. Range : 2.15 – 1.83 = 0.32   The range is 0.32 m.

9. Average : (70 x 25 + 60 x 30) ÷ 55 = (1750 + 1800) ÷ 55 = 64.5
The average percentage is 64.55.

10. Sum : 20 x 3 – 15 = 45   The sum of the other 2 numbers is 45.

11. a. Mean: (8 + 8 + 9 + 9 + 12 + 13 + 15 + 17 + 31) ÷ 9 = 13.56
    The mean is 13.56; the median is 12; the modes are 8 and 9.
    b. The mean best reflects the data because it is the average of all the data; the modes only show the most common values.
    c. I would use the mean to show how bad the problem was.
    d. I would use the mode since it could make the problem look less serious.

12. a. Average : (750 + 900 + 1000 + 1050 + 1100) ÷ 5 = 960
    The average sales are 960 million dollars ($960 000 000).
    b. This impression is created because the vertical scale does not start at zero, so it makes the increase look greater.

13. a. 104.68/sq. km;   334.22/sq. km;   225.59/sq. km;
    183.33/sq. km;   453.61/sq. km
    b. South Korea is the most densely populated country.
    c. Mean : (197 + 125 + 67 + 22 + 44) ÷ 5 = 91
    The mean population size is 91 millions (91 000 000).
    d. Put the data in order : 22, 44, 67, 125, 197
    The median population size is 67 millions.
    e. Bar graph. Here, 2 bar graphs would be appropriate --- one for population and one for area, or we may use 1 bar graph for population density.

Challenge

5 of the 7 test scores should be 34, 60, 70, 70 and 71.
The sum of the other 2 tests : 55 x 7 – (34 + 60 + 70 + 70 + 71) = 80
A. 34, 39, 41, 60, 70, 70, 71
B. 34, 38, 42, 60, 70, 70, 71
C. 34, 37, 43, 60, 70, 70, 71
D. 34, 36, 44, 60, 70, 70, 71
E. 34, 35, 45, 60, 70, 70, 71

ISBN: 978-1-897164-21-1

Unit 7

1. a. $\frac{24}{50}$ (or $\frac{12}{25}$) of the flips turned up heads.

   b. There is no definite answer.

   c. She should expect heads to come up 250 times.

   d. $\frac{1}{2}$ of the flips should come up heads.

   e. The probability of getting heads is $\frac{1}{2}$.

   f. Yes, it is possible but it is unlikely to happen.

2.
   H — H (H H) Carol
       — T (H T) Debbie

   T — H (T H) Debbie
       — T (T T) Carol

   They are both likely to win because their chances of winning are the same.

3.
   H — H — H (H H H) Eric
         — T (H H T) Frank
     — T — H (H T H) Frank
         — T (H T T) Frank

   T — H — H (T H H) Frank
         — T (T H T) Frank
     — T — H (T T H) Frank
         — T (T T T) Eric

   Frank is likely to win.

4. a. Spinner A : P (3) = $\frac{1}{3}$

   Spinner B : P (3) = $\frac{1}{2}$

   Spinner C : P (3) = $\frac{1}{6}$

   b. Spinner A : He should expect 10 times.
   Spinner B : He should expect 15 times.
   Spinner C : He should expect 5 times.

   c. No, it doesn't depend on the number of spins.

   d. Spinner A.

5. a. It doesn't matter because the priority of guessing does not affect the probability of getting the right coloured ball.

   b. P (W) = $\frac{20}{60}$ = $\frac{1}{3}$ The probability is $\frac{1}{3}$.

   c. The most likely outcome is getting a black ball.

6. a. P (4) = $\frac{1}{6}$ The probability is $\frac{1}{6}$.

   b. P (5) = $\frac{1}{6}$; P (6) = $\frac{1}{6}$; P (>4) = $\frac{1}{6}$ + $\frac{1}{6}$ = $\frac{2}{6}$ = $\frac{1}{3}$

   The probability is $\frac{1}{3}$.

7. a. P (H) = $\frac{2}{8}$ or $\frac{1}{4}$

   The probability that there will be hamburgers for lunch is $\frac{1}{4}$.

   b. I would like her to use spinner B because the chance of getting spaghetti in spinner B is higher than that in spinner A.

   c. He should get 4 points because the probability of getting lasagna is half the probability of getting a hamburger.

8. a.

⊕	1	2	3	4	5	6
1	2	3	4	5	6	7
2	3	4	5	6	7	8
3	4	5	6	7	8	9
4	5	6	7	8	9	10
5	6	7	8	9	10	11
6	7	8	9	10	11	12

   b. No, they are not.

   c. P (4) = $\frac{3}{36}$ = $\frac{1}{12}$ The probability is $\frac{1}{12}$.

   d. P (7) = $\frac{6}{36}$ = $\frac{1}{6}$ 7 is most likely.

9. P (a double) = $\frac{6}{36}$ = $\frac{1}{6}$
   The probability of getting a double is $\frac{1}{6}$.

10. a. 48 different outcomes can only be obtained by rolling 2 dice with either 4 and 12 faces or 6 and 8 faces.
    Possible outcomes of 2 dice (the sum of 6) with 4 and 12 faces:
    (1, 5), (2, 4), (4, 2) and (3, 3)
    Possible outcomes of 2 dice (the sum of 6) with 6 and 8 faces:
    (1, 5), (5, 1), (2, 4), (4, 2) and (3, 3)
    Trish is using the dice with 6 and 8 faces.

    b. No. of possible outcomes : 48;
    Favourable outcomes : (6, 7), (5, 8)
    The probability is $\frac{2}{48}$ ($\frac{1}{24}$).

11. a. (Quebec, Montreal, Hull); (Quebec, Hull, Montreal);
    (Montreal, Quebec, Hull); (Montreal, Hull, Quebec);
    (Hull, Quebec, Montreal); (Hull, Montreal, Quebec).
    There are 6 different orders.

    b. No. It depends on where you start.

12. a. P (6) = $\frac{1}{6}$ x $\frac{1}{6}$ x $\frac{1}{6}$ = $\frac{1}{216}$ The probability is $\frac{1}{216}$.

    b. P (4) = $\frac{1}{216}$; P (5) = $\frac{1}{216}$

    P (4 or 5) = $\frac{1}{216}$ + $\frac{1}{216}$ = $\frac{2}{216}$ ($\frac{1}{108}$)

    The probability is $\frac{1}{108}$.

    c. P (all match) = $\frac{6}{216}$ ($\frac{1}{36}$)   The probability is $\frac{1}{36}$.

13. P (6) = $\frac{1}{6}$ 10 times out of 60 will win.
    Money I win : 2 x 10 = 20
    I expect to win $20.00 by rolling the die 60 times.

14. a. All possible outcomes : {(1,2);(1,3);(1,4);(1,5);(2,1);(2,3);(2,4);(2,5);   (3,1);(3,2);(3,4);(3,5);(4,1);(4,2);(4,3);(4,5);(5,1);(5,2);(5,3);(5,4)}
    The probability of getting both of them even is $\frac{2}{20}$ ($\frac{1}{10}$).

    b. No. of favourable outcomes : 12;
    No. of possible outcomes : 20
    The probability of getting one of them even is $\frac{12}{20}$ ($\frac{3}{5}$).

15. Possible outcomes : {AA; AB; AC; AB; AC; BC}; P (W) = $\frac{1}{6}$
    The probability that you will win is $\frac{1}{6}$.

Challenge

4CD = A, B, C, D.
Possible outcomes : {(A,B); (A,C); (A,D); (B,C); (B,D); (C, D)}
There are 6 different ways.

Final Review

1. D	2. C	3. B	4. D	5. C
6. B	7. C	8. B	9. A	10. D

11. Mean : (85 + 82 + 91 + 72 + 75 + 85) ÷ 6 = 81.67
    Median : (82 + 85) ÷ 2 = 83.5   Mode : 85
    Julie should choose the mode because it shows the highest score.

12. There are 2 possible ways :
    • To score 85 or more on the 3 tests
    • To score less than 85 on 1 test and 85 or more on the other 2 tests

13. Sum of the 3 upcoming tests :
    (80 x 9) – (85 + 82 + 91 + 72 + 75 + 85) = 230
    The sum of the scores on the 3 upcoming tests should be 230, e.g. 75, 75, 80.

14. A bar graph should be used because it shows the actual score on each test clearly.

15. There are two scores (72 and 75) equivalent to Grade B.
    P(70 – 79%) = $\frac{2}{6}$ = $\frac{1}{3}$ The probability is $\frac{1}{3}$.

16. The median of Julie's tests is 83.5. Mary should get 83 or below on 2 other tests, 84 on 1 test and 84 or above on the remaining tests.
    (e.g. 81, 82, 83, *84, 85, 86   *2nd test)

17. She should get more scores of 85 than other scores on her first 6 Math tests.

18. a. No. of hours : $\frac{30}{360}$ x 24 = 2

A typical 12-year-old child spends 2 hours per day on homework.

b. Sleeping : $\frac{120}{360}$ x 100% = 33.3%

About 33.3% of a day is spent on sleeping.

c. It shows how the day is divided up.

d. Percentage on school and homework might increase. Percentage on eating and sleeping might decrease.

19. a. Total no. of chocolate bars : 500 x 20 = 10000

Total no. of coupons : $\frac{10000}{50}$ + $\frac{10000}{200}$ = 200 + 50 = 250

Fair distribution : $\frac{10000}{250}$ = 40

Every 40 bars (2 boxes) should contain 1 of the 2 coupons.

b. P (win) = $\frac{10}{40}$ = $\frac{1}{4}$ The probability is $\frac{1}{4}$.

c. Cost : 200 x 0.8 x 0.5 + 50 x 0.8 x 2 = 80 + 80 = 160
The total cost of the promotion is $160.00.

20. a.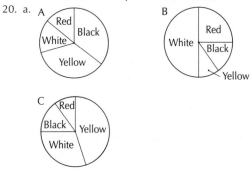

The proportion of the colours on each spinner is based on the number of times they occured when Jane spun each of them 100 times.

b.
H < H – H H
        T – H T

T < H – T H
        T – T T

P (H H) = $\frac{1}{4}$;   No. of times : 20 x $\frac{1}{4}$ = 5

He should expect to win 5 times.

21. a.

tens	ones
13	5, 8, 8, 9, 9
14	0, 0, 0, 1, 2, 4, 4, 5, 6, 8, 9
15	0, 0, 0, 0, 1, 3, 6, 8, 9

The median age is 145 months.

b.   Ages of Grade 7 Students

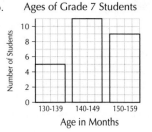

22. a. JCBD, JCDB, JBCD, JBDC, JDCB, JDBC,
CJBD, CJDB, CBJD, CBDJ, CDJB, CDBJ,
BJCD, BJDC, BCJD, BCDJ, BDJC, BDCJ,
DBJC, DBCJ, DJBC, DJCB, DCBJ, DCJB

b. There are 24 different running orders.

c. P(J) = $\frac{6}{24}$ = $\frac{1}{4}$ The probability is $\frac{1}{4}$.

d. There are 6 different running orders.

ISBN: 978-1-897164-21-1